Tucholsky Wagner Zola Scott Sydow Freud Schlegel
Turgenev Wallace Fonatne
Twain Walther von der Vogelweide Fouqué Friedrich II. von Preußen
Weber Freiligrath Frey
Fechner Fichte Weiße Rose von Fallersleben Kant Ernst Frommel
Richthofen
Engels Fielding Hölderlin Tacitus Dumas
Fehrs Faber Flaubert Eichendorff
Eliasberg Ebner Eschenbach
Feuerbach Maximilian I. von Habsburg Fock Zweig
Ewald Eliot Vergil
Goethe Elisabeth von Österreich London
Mendelssohn Balzac Shakespeare Dostojewski Ganghofer
Lichtenberg Rathenau Doyle Gjellerup
Trackl Stevenson Hambruch
Mommsen Tolstoi Lenz Droste-Hülshoff
Thoma von Arnim Hanrieder
Dach Verne Hägele Hauff Humboldt
Karrillon Reuter Rousseau Hagen Hauptmann Gautier
Garschin Defoe Baudelaire
Damaschke Hebbel
Descartes Hegel Kussmaul Herder
Wolfram von Eschenbach Schopenhauer Rilke George
Bronner Darwin Dickens Grimm Jerome
Melville Bebel Proust
Campe Horváth Aristoteles Voltaire Federer Herodot
Bismarck Vigny Barlach Heine
Gengenbach
Storm Casanova Tersteegen Grillparzer Georgy
Chamberlain Lessing Langbein Gilm Gryphius
Brentano Lafontaine
Claudius Schiller Kralik Iffland Sokrates
Strachwitz Schilling
Katharina II. von Rußland Bellamy Raabe Gibbon Tschechow
Gerstäcker
Löns Hesse Hoffmann Gogol Wilde Vulpius
Gleim
Luther Heym Hofmannsthal Morgenstern
Klee Hölty Goedicke
Roth Heyse Klopstock Kleist
Luxemburg Puschkin Homer Mörike Musil
La Roche Horaz
Machiavelli Kierkegaard Kraft Kraus
Navarra Aurel Musset Moltke
Lamprecht Kind Kirchhoff Hugo
Nestroy Marie de France
Laotse Ipsen Liebknecht
Nietzsche Nansen Ringelnatz
Marx Lassalle Gorki Klett Leibniz
von Ossietzky May vom Stein Lawrence Irving
Petalozzi Knigge
Platon Michelangelo Kock Kafka
Sachs Poe Liebermann
Korolenko
de Sade Praetorius Mistral Zetkin

The publishing house tredition has created the series **TREDITION CLASSICS**. It contains classical literature works from over two thousand years. Most of these titles have been out of print and off the bookstore shelves for decades.

The book series is intended to preserve the cultural legacy and to promote the timeless works of classical literature. As a reader of a **TREDITION CLASSICS** book, the reader supports the mission to save many of the amazing works of world literature from oblivion.

The symbol of **TREDITION CLASSICS** is Johannes Gutenberg (1400 – 1468), the inventor of movable type printing.

With the series, tredition intends to make thousands of international literature classics available in printed format again – worldwide.

All books are available at book retailers worldwide in paperback and in hardcover. For more information please visit: www.tredition.com

tredition was established in 2006 by Sandra Latusseck and Soenke Schulz. Based in Hamburg, Germany, tredition offers publishing solutions to authors and publishing houses, combined with worldwide distribution of printed and digital book content. tredition is uniquely positioned to enable authors and publishing houses to create books on their own terms and without conventional manufacturing risks.

For more information please visit: www.tredition.com

# Vegetable Dyes Being a Book of Recipes and Other Information Useful to the Dyer

Ethel M. Mairet

# Imprint

This book is part of the TREDITION CLASSICS series.

Author: Ethel M. Mairet
Cover design: toepferschumann, Berlin (Germany)

Publisher: tredition GmbH, Hamburg (Germany)
ISBN: 978-3-8491-8616-6

www.tredition.com
www.tredition.de

# CONTENTS



[Pg 1]

# CHAPTER I

# WOOL SILK COTTON AND LINEN

WOOLS are of various kinds: —

*Highland, Welsh and Irish* wools are from small sheep, not far removed from the wild state, with irregular short stapled fleeces.

*Forest or Mountain sheep* (Herdwick, Exmoor, Cheviot, Blackfaced, Limestone) have better wool, especially the Cheviot, which is very thick and good for milling.

*Ancient Upland,* such as South Down, are smaller sheep than the last named, but the wool is softer and finer.

*Long Woolled sheep,* (Lincolns, Leicester) with long staple wool (record length, 36".) and fleeces weighing up to 12 lbs. The Leicester fleece is softer, finer and better than Lincoln.

To the end of the 18th century *Spanish wool* was the finest and best wool in the world. Spanish sheep have since been introduced into various countries, such as Saxony, Australia, Cape Colony, New Zealand; and some of the best wools now come from the Colonies.

*Alpaca, Vicuna and Llama* wools are from different species of American goats.

*Mohair* from the Angora goat of Asia Minor.

*Kashmir Wool* from the Thibetan goat.

*Camel* hair, the soft under wool of the camel, which is shed annually.

The colour of wool varies from white to a very dark brown black, with all shades of fawn, grey and brown in between. The natural colours are not absolutely fast to light but tend to bleach slightly with the sun.

The principal fleeces are:

*Lambs,* 3 to 6 months growth, the finest, softest and most elastic wool. [Pg 2]

*Hogs and Tegs*: the first shearing of sheep that have not been shorn as lambs.

*Wethers*: all clips succeeding the first shearing.

Wool comes into the market in the following condition. 1. *In the grease*, not having been washed and containing all the impurities. 2. *Washed*, with some of the grease removed and fairly clean. 3. *Scoured*, thoroughly cleaned and all grease removed.

Wool can be dyed either in the fleece, in the yarn, or in the woven cloth. Raw wool always contains a certain amount of natural grease. This should not be washed out until it is ready for dyeing, as the grease keeps the moth out to a considerable extent. Hand spun wool is generally spun in the oil to facilitate spinning. All grease and oil must be scoured out before dyeing is begun, and this must be done very thoroughly or the wool will not take the colour.

## WATER

A constant supply of clean soft water is an absolute necessity for the dyer. Rain water should be collected as much as possible, as this is the best water to use. The dye house should be by a river or stream, so that the dyer can wash with a continuous supply. Spring and well water is, as a rule, hard, and should be avoided. In washing, as well as in dyeing, hard water is injurious for wool. It ruins the brilliancy of the colour, and prevents the dyeing of some colours. Temporary hardness can be overcome by boiling the water (20 to 30 minutes) before using. An old method of purifying water, which is still used by some silk and wool scourers, is to boil the water with a little soap, skimming off the surface as it boils. In many cases it is sufficient to add a little acetic acid to the water.

## TO WASH WOOL

In a bath containing 10 gallons of warm water add 4 fluid ounces of ammonia fort, .880, 1 lb. soda, and 2 oz. soft soap, (potash soap). Stir well until all is dissolved. Dip the wool in [Pg 3] and leave for 2 minutes, then squeeze gently and wash in warm water until quite clear.

*Or* to 10 gallons of water add 6 oz. ammonia and 3 oz. soft soap. The water should never be above 140°F. and all the washing water should be of about the same temperature.

Fleece may be washed in the same way, but great care should be taken not to felt the wool—the less squeezing the better.

There are four principal methods of dyeing wool.

1st.—The wool is boiled first with the mordant and then in a fresh bath with the dye.

2nd.—The wool is boiled first with the dye, and when it has absorbed as much of the colour as possible the mordant is added to the same bath, thus fixing the colour.

A separate bath can be used for each of these processes, in which case each bath can be replenished and used again for a fresh lot of wool.

3rd.—The wool is boiled with the mordant and dye in the same bath together. The colour, as a rule, is not so fast and good as with a separate bath, though with some dyes a brighter colour is obtained.

4th.—The wool is mordanted, then dyed, then mordanted again. This method is adopted to ensure an extremely fast colour. The mordant should be used rather sparingly.

## SILK

There are two kinds of silk (1) *raw silk* (reeled silk, thrown silk, drawn silk), and (2) *waste silk* or spun silk.

Raw silk is that directly taken from the cocoons. Waste silk is the silk from cocoons that are damaged in some way so that they cannot be reeled off direct. It is, therefore, carded and spun, like wool or cotton.

Silk in the raw state is covered with a silk gum which must be boiled off before dyeing is begun. It is tied up in canvas bags and boiled up in a strong solution of soap for three or four hours until all the gum is boiled off. If it is a yellow gum, the silk is wrought first in a solution of soft soap [Pg 4] at a temperature just below boiling point for about an hour, then put into bags and boiled. After boiling, the soap is well washed out.

Generally speaking, the affinity of silk for dyes is similar but weaker in character to that of wool. The general method for dyeing is the same as for wool, except, in most cases, lower temperatures are used in the mordanting. In some cases, soaking in a cold concentrated solution of the mordant is sufficient. The dyeing of some colours is also at low temperature.

## COTTON

Cotton is the down surrounding the seeds in pods of certain shrubs and trees growing in tropical and semi-tropical countries. First introduced into Europe by the Saracens, it was manufactured into cloth in Spain in the early 13th century. Cotton cloth was first made in England in the early 17th century.

The colour of cotton varies from deep yellow to white. The fibre differs in length, the long stapled being the most valued. It is difficult to dye and requires a special preparation.

A few of the natural dye stuffs are capable of dyeing cotton direct, without a mordant, such as Turmeric, Barberry bark, safflower, annatto. For other dyes cotton has a special attraction, such as catechu.

## LINEN

Linen is flax, derived from the decomposed stalks of a plant of the genus Linum.

It grows chiefly in Russia, Belgium, France, Holland and Ireland. The plants after being gathered are subjected to a process called "retting" which separates the fibre from the decaying part of the plant. In Ireland and Russia this is usually done in stagnant water, producing a dark coloured flax. In Belgium, Holland, and France, retting is carried out in running water, and the resulting flax is a lighter colour. Linen is more difficult to dye than cotton, probably on account of the hard nature of the fibre. The same processes are used for dyeing linen as for cotton. [Pg 5]

*To Bleach Linen* — (For 13 to 15 yards linen). Boil 1/2 lb. soap and 1/2 lb. soda in a gallon of water. Put it in a copper and fill up with water, leaving room for the linen to be put in. Put in the linen and bring to the boil. Boil for 2 hours, keeping it under the water and covered. Stir occasionally. Then spread out on the grass for 3 days, watering it when it gets dry. Repeat this boiling and grassing 3 weeks. The linen is then pure white.

*To bleach linen a cream colour* — Boil 1/2 lb. soap and 1/2 lb. soda in a gallon of water. Fill copper up with water and put in linen. Boil for 2 hours. Repeat this once a day for 4 days. The linen should not be wrung out but kept in the water till ready to be put into the fresh bath.

[Pg 6]

# CHAPTER II

# MORDANTS

Any dye belongs to one of two classes. *Substantive*, giving colouring directly to the material: and *adjective*, which includes the greater number of dyes and requires the use of a mordant to bring out the colour.

There are thus two processes concerned with the dyeing of most colours; the first is mordanting and the second is the colouring or actual dyeing. The mordanting prepares the stuff to receive the dye (*mordere, to bite*).

The early French dyers thought that a mordant had the effect of opening the pores of the fibre, so that the dye could more easily enter; but according to Hummel, and later dyers, the action of the mordant is purely chemical; and he gives a definition of a mordant as "the body, whatever it may be, which is fixed on the fibre in combination with any given colouring matter." The mordant is first precipitated on to the fibre and combines with the colouring matter in the subsequent dye bath. But, whether the action is chemical or merely physical, the fact remains that all adjective dyes need this preparation of the fibre before they will fix themselves on it. The use of a mordant, though not a necessity, is sometimes an advantage when using substantive dyes.

In early days the leaves and roots of certain plants were used. This is the case even now in India and other places where primitive dyeing methods are still carried on. Alum has been known for centuries in Europe. Iron and tin filings have also been used. Alum and copperas have been known in the Highlands long ages.

*Mordants* should not affect the physical characteristics of the fibres. Sufficient time must be allowed for the mordant to penetrate the fibre thoroughly. If the mordant is only superficial, the dye will be uneven: it will fade and will not be as brilliant as it should be. The brilliancy and fastness of [Pg 7] Eastern dyes are probably due to a great extent to the length of time taken over the various processes of dyeing. *The longer time that can be given to each process, the more satisfactory will be the result.*

Different mordants give different colours with the same dye stuff. For example:—Cochineal, if mordanted with alum, will give a crimson colour; with iron, purple; with tin, scarlet; and with chrome or copper, purple. Logwood, also, if mordanted with alum, gives a mauve colour; if mordanted with chrome, it gives a blue. Fustic, weld, and most of the yellow dyes, give a greeny yellow with alum, but an old gold colour with chrome; and fawns of various shades with other mordants.

Silk and wool require very much the same preparation except that in the case of silk, high temperatures should be avoided. Wool is generally boiled in a weak solution of whatever mordant is used. With silk, as a rule, it is better to use a cold solution, or a solution at a temperature below boiling point. Cotton and linen are more difficult to dye than wool or silk. Their fibre is not so porous and will not hold the dye stuff without a more complicated preparation. The usual method of preparing linen or cotton is to boil it first with some astringent. The use of astringents in dyeing depends upon the tannic acid they contain. In combination with ordinary mordants, tannic acid aids the attraction of the colouring matter to the fibre and adds brilliancy to the colours. The astringents mostly used are tannic acid, gall nuts, sumach and myrobalams. Cotton has a natural attraction for tannic acid, so that when once steeped in its solution it is not easily removed by washing.

## ALUM

This is the most generally used of all the mordants, and has been known as such from early times in many parts of the world. For most colours a certain proportion of cream of tartar should be added to the alum bath as it helps to [Pg 8] brighten the ultimate colour. The usual amount of alum is a quarter of a pound to a pound of wool. As a rule, less mordant is needed for light colours than for dark. Excess of alum is apt to make the wool sticky. The usual length of time for boiling is about an hour. Some dyers give as much as 2-1/2 hours.

*Example of mordanting with alum* — 1/4 lb. of Alum and 1 oz. cream of tartar for every pound of wool. This is dissolved and when the water is warm the wool is entered. Raise to boiling point and boil for one hour. The bath is then taken off the fire and allowed to cool over night. The wool is then wrung out (not washed) and put away in a linen bag in a cool place for 4 or 5 days, when it is ready for dyeing, after being thoroughly washed.

## IRON

*(Ferrous Sulphate, copperas, green vitriol.)*

Iron is one of the oldest mordants known and is largely used in wool and cotton dyeing. It is almost as important as alum. The temperature of the mordanting bath must be raised very gradually to boiling point or the wool will dye unevenly. A general method of dealing with copperas is to boil the wool first in a decoction of the colouring matter and then add the mordant to the same bath in a proportion of 5 to 8 per cent of the weight of the wool, and continue boiling for half an hour or so longer. With some dyes a separate bath is needed, such as with Camwood or Catechu. Great care is needed in the using of copperas, as, unless it is thoroughly dissolved and mixed with the water before the wool is entered, it is apt to stain the wool. It also hardens wool if used in excess or if boiled too long. A separate bath should always be kept for dyes or mordants containing iron. The least trace of it will dull colours and it will spoil the brilliancy of reds, yellows and oranges.

Copperas is mostly used for the fixing of wool colours [Pg 9] (Fustic, etc.) to produce brown shades; the wool being boiled first in a decoction of the dye for about 1 hour, and then for 1/2 an hour with the addition of 5 to 8 per cent of copperas. If used for darkening colours, copperas is added to the bath after the dyeing, and the boiling continued for 15 to 20 mins.

## TIN

*(Stannous chloride, tin crystals, tin salts, muriate of tin.)*

Tin is not so useful as a mordant in itself, but as a modifying agent with other mordants. It must always be used with great care, as it tends to harden the wool, making it harsh and brittle. Its general effect is to give brighter, clearer and faster colours than the other mordants. When used as a mordant before dyeing, the wool is entered into the *cold* mordant bath, containing 4 per cent of stannous chloride and 2 per cent oxalic acid; the temperature is gradually raised to boiling, and kept at this temperature for 1 hour. It is sometimes added to the dye bath towards the end of dyeing, to intensify and brighten the colour. It is also used with cochineal for scarlet on wool in the one bath method.

## CHROME

*(Potassium dichromate. Bichromate of Potash.)*

Chrome is a modern mordant, unknown to the dyer of fifty years ago. It is excellent for wool and is easy to use and very effective in its action. Its great advantage is that it leaves the wool soft to the touch, whereas the other mordants are apt to harden the wool. The wool should be boiled for 1 to 1-1/2 hours with bichromate of potash in the proportion of 2 to 4 per cent of the wool. It is then washed well and immediately dyed. Wool mordanted with chrome should not be exposed to light, but should be kept well covered with the liquid while being mordanted, else it is liable to dye unevenly. An excess of chrome impairs the colour, 3 per cent of chrome is a safe quantity to use for ordinary dyeing. It should be dissolved in the bath while the water is heating. The wool is entered and the bath [Pg 10] gradually raised to the boiling point, and boiled for 3/4 of an hour.

## COPPER

*(Copper Sulphate, Verdigris, Blue Vitriol, Blue Copperas, Bluestone.)*

Copper is rarely used as a mordant. It is usually applied as a saddening agent, that is, the wool is dyed first, and the mordant applied afterwards to fix the colour. With *cream of tartar* it is used sometimes as an ordinary mordant before dyeing, but the colours so produced have no advantage over colours mordanted by easier methods.

[Pg 11]

# CHAPTER III

# BRITISH DYE PLANTS

On the introduction of foreign dye woods and other dyes during the 17th and 18th centuries, the native dye plants were rapidly displaced, except in some out of the way places such as the Highlands and parts of Ireland. Some of these British dye plants had been used from early historical times for dyeing. Some few are still in use in commercial dye work (pear, sloe, and a few others); but their disuse was practically completed during the 19th century, when the chemical dyes ousted them from the market.

The majority of these plants are not very important as dyes, and could not probably now be collected in sufficient quantities. Some few, however, are important, such as woad, weld, heather, walnut, alder, oak, some lichens; and many of the less important ones would produce valuable colours if experiments were made with the right mordants. Those which have been in use in the Highlands are most of them good dyes. Among these are Ladies Bedstraw, whortleberry, yellow iris, bracken, bramble, meadow sweet, alder, heather and many others. The yellow dyes are most plentiful and many of these are good fast colours. Practically no good red, in quantity, is obtainable. Madder is the only reliable red dye among plants, and that is no longer indigenous in England. Most of the dye plants require a preparation of the material to be dyed, with alum, or some other mordant, but a few, such as Barbary and some of the lichens, are substantive dyes, and require no mordant.

## PLANTS WHICH DYE RED

Birch. *Betula alba.* Fresh inner bark.

Bed-straw. *Gallium boreale.* Roots.

Common Sorrel. *Rumex acetosa.* Roots. [Pg 12]

Dyer's Woodruff. *Asperula tinctoria.* Roots.

Evergreen Alkanet. *Anchusa sempervirens.*

Gromwell. *Lithospermum arvense.*

Lady's Bedstraw. *Gallium verum.* Roots.

Marsh Potentil. *Potentilla Comarum.* Roots.

Potentil. *Potentilla Tormentilla.* Roots.

Wild Madder. *Rubia peregrina.*

## PLANTS WHICH DYE BLUE

Devil's Bit. *Scabiosa succisa.* Leaves prepared like woad.

Dog's Mercury. *Mercurialis perennis.*

Elder. *Sambucus nigra.* Berries.

Privet. *Ligustrum vulgare.* Berries with alum and salt.

Red bearberry. *Arctostaphylos Uva-Ursi.*

Sloe. [A] *Prunus communis.* Fruit.

Whortleberry or Blaeberry. *Vaccinium Myrtillus.* Berries.

Woad. *Isatis tinctoria.*

Yellow Iris. *Iris Pseudacorus.* Roots.

## PLANTS WHICH DYE YELLOW

Agrimony. *Agrimonia Eupatoria.*

Ash. *Fraxinus excelsior.* Fresh inner bark.

Barberry. *Berberis vulgaris.* Stem and root.

Birch. Leaves.

Bog Asphodel. *Narthecium ossifragum.*

Bog Myrtle or Sweet Gale. *Myrica Gale.*

Bracken. *Pteris aquilina.* Roots. Also young tops.

Bramble. *Rubus fructicosus.*

Broom. *Sarothammus Scoparius.*

Buckthorn. *Rhamnus frangula* and *R. cathartica.* Berries and Bark.

Common dock. *Rumex obtusifolius.* Root.

Crab Apple. *Pyrus Malus.* Fresh inner bark.

Dyer's Greenwood. *Genista tinctoria.* Young shoots and leaves.

Gorse. *Ulex Europæus.* Bark, flowers and young shoots.

[Pg 13]

Heath. *Erica vulgaris.* With Alum.

Hedge stachys. *Stachys palustris.*

Hop. *Humulus lupulus.*

Hornbeam. *Carpinus Betulus.* Bark.

Kidney Vetch. *Anthyllis Vulnararia.*

Ling. *Caluna vulgaris.*

Marsh Marigold. *Caltha palustris.*

Marsh potentil. *Potentilla Comarum.*

Meadow Rue. *Thalictrum flavum.*

Nettle. *Urtica.* With Alum.

Pear. Leaves.

Plum. "

Polygonum Hydropiper.

Polygonum Persecaria.

Poplar. Leaves.

Privet. *Ligustrum vulgare.* Leaves.

S. John's Wort. *Hypericum perforatum.*

Sawwort. [B] *Serratula tinctoria.*

Spindle tree. *Euonymus Europæus.*

Stinking Willy, or Ragweed. *Senecio Jacobæa.*

Sundew. *Drosera.*

Teasel. *Dipsacus Sylvestris.*

Way-faring tree. *Viburnum lantana.* Leaves.

Weld. *Reseda luteola.*

Willow. [C] Leaves.

Yellow Camomile. *Anthemis tinctoria.*

Yellow Centaury. *Chlora perfoliata.*

Yellow Corydal. *Corydalis lutea.*

[Pg 14]

## PLANTS WHICH DYE GREEN

Elder. *Sambucus nigra.* Leaves with alum.

Flowering reed. *Phragmites communis.* Flowering tops, with copperas.

Larch. Bark, with alum.

Lily of the valley. *Convalaria majalis.* Leaves.

Nettle. *Urtica dioica* and *U. Urens.*

Privet. *Ligustrum vulgare.* Berries and leaves, with alum.

## PLANTS WHICH DYE BROWN

Alder. *Alnus glutinosa.* Bark.

Birch. *Betula alba.* Bark.

Hop. *Humulus lupulus.* Stalks give a brownish red colour.

Onion. Skins.

Larch. Pine needles, collected in Autumn.

Oak. *Quercus Robur.* Bark.

Red currants, with alum.

Walnut. Root and green husks of nut.

Water Lily. *Nymphæa alba.* Root.

Whortleberry. *Vaccinium Myrtillus.* Young shoots, with nut galls.

Dulse. (Seaweed.)

Lichens.

# PLANTS WHICH DYE PURPLE

Byrony. *Byronia dioica.* Berries.

Damson. Fruit, with alum.

Dandelion. *Taraxacum Dens-leonis.* Roots.

Danewort. *Sambucus Ebulus.* Berries.

Deadly nightshade. *Atropa Belladonna.*

Elder. *Sambucus nigra.* Berries, with alum, a violet; with alum and salt, a lilac colour.

Sundew. *Drosera.*

Whortleberry or blaeberry. *Vaccinium myrtillus.* It contains a blue or purple dye which will dye wool and silk without mordant.

[Pg 15]

# PLANTS WHICH DYE BLACK

Alder. *Alnus glutinosa.* Bark, with copperas.

Blackberry. *Rubus fruticosus.* Young shoots, with salts of iron.

Dock. *Rumex.* Root.

Elder. Bark, with copperas.

Iris. *Iris Pseudacorus.* Root.

Meadowsweet. *Spirea Ulmaria.*

Oak. Bark and acorns.

## FOOTNOTES:

[A] "On boiling sloes, their juice becomes red, and the red dye which it imparts to linen changes, when washed with soap, into a bluish colour, which is permanent."

[B] "Sawwort, which grows abundantly in meadows, affords a very fine pure yellow with alum mordant, which greatly resembles weld yellow. It is extremely permanent."

[C] "The leaves of the sweet willow, *salix pentandra*, gathered at the end of August and dried in the shade, afford, if boiled with about one thirtieth potash, a fine yellow colour to wool, silk and thread, with alum basis. All the 5 species of Erica or heath growing on this island are capable of affording yellow much like those from the dyer's broom; also the bark and shoots of the Lombardy poplar, *populus pyramidalis*. The three leaved hellebore, *helleborus trifolius*, for dyeing wood yellow, is used in Canada. The seeds of the purple trefoil, lucerne, and fenugreek, the flowers of the French marigold, the camomile, *antemis tinctoria*, the ash, *fraxinus excelsior*, fumitory, *fumaria officinalis*, dye wool yellow." "The American golden rod, *solidago canadensis*, affords a very beautiful yellow to wool, silk and cotton upon an aluminous basis." *Bancroft*.

[Pg 16]

# CHAPTER IV

# THE LICHEN DYES

Some of the most useful dyes and the least known are to be found among the Lichens. They seem to have been used among peasant dyers from remote ages, but apparently none of the great French dyers used them, nor are they mentioned in any of the old books on dyeing. The only Lichen dyes that are known generally among dyers are Orchil and Cudbear, and these are preparations of lichens, not the lichens themselves. They are still used in some quantity and are prepared rather elaborately. But a great many of the ordinary lichens yield very good and permanent dyes. The *Parmelia saxatilis* and *Parmelia omphalodes*, are largely used in the Highlands and West Ireland, for dyeing brown of all shades. No mordant is needed, and the colours produced are the fastest known. "Crottle" is the general name for Lichens in Scotland. They are gathered off the rocks in July and August, dried in the sun, and used to dye wool, without any preparation. The crottle is put into the bath with a sufficient quantity of water, boiled up, allowed to cool, then boiled up with the wool until the shade required is got. This may take from one to three or four hours, as the dye is not rapidly taken up by the wool. Other dyers use it in the following way: A layer of crottle, a layer of wool, and so on until the bath is full; fill with cold water and bring to the boil, and boil till the colour is deep enough. The wool does not seem to be affected by keeping it in the dye a long time. A small quantity of acetic acid put with the Lichen is said to assist in exhausting the colour.

The grey Lichen, *Ramalina scopulorum* dyes a fine shade of yellow brown. It grows very plentifully on old stone walls, especially by the sea, and in damp woods, on trees, and on old rotten wood. Boil the Lichen up in sufficient water one day, and the next put in the wool, and boil up again till [Pg 17] the right colour is got. If the wool is left in the dye for a day or more after boiling it absorbs more colour, and it does not hurt the wool but leaves it soft and silky to the touch, though apt to be uneven in colour. Some mordant the wool first with alum, but it does not seem to need it.

The best known of the dye Lichens are *Parmelia saxatilis* and *Parmelia omphalodes* which are still largely used in Scotland and Ireland for dyeing wool for tweeds. The well-known Harris tweed smell is partly due to the use of this dye.

Other Lichens also known for their dyeing properties are: *Parmelia caperata*, or Stone Crottle, which contains a yellow dye, *P. ceratophylla*, or Dark Crottle, and *P. parietina*, the common wall Lichen, which gives a colour similar to that of the Lichen itself, yellowish brown. A deep red colour can be got from the dull grey friable Lichen, common on old stone walls. The bright yellow Lichen, growing on rocks and walls, and old roofs, dyes a fine plum colour, if the wool is mordanted first with Bichromate of Potash.

In Sweden, Scotland and other countries the peasantry use a Lichen, called *Lecanora tartarea* to furnish a red or crimson dye. It is found abundantly on almost all rocks, and also grows on dry moors. It is collected in May and June, and steeped in stale urine for about three weeks, being kept at a moderate heat all the time. The substance having then a thick and strong texture, like bread, and being of a blueish black colour, is taken out and made into small cakes of about 3/4 lb. in weight, which are wrapped in dock leaves and hung up to dry in peat smoke. When dry it may be preserved fit for use for many years; when wanted for dyeing it is partially dissolved in warm water; 5 lbs. of Korkalett is considered sufficient for about 4 Scotch ells of cloth. The colour produced is a light red. It is used in the dyeing of yarn as well as of cloth.

In Shetland, the *Parmelia saxatilis* (Scrottyie) is used to dye brown. It is found in abundance on argillaceous rocks. It is considered best if gathered late in the year, and is generally collected in August.

Linnaeus mentions that a beautiful red colour may be prepared from the Lichen *Gyrophora pustulata*. *G. Cylindrica* [Pg 18] is used by Icelanders for dyeing woollen stuffs a brownish green colour. In Sweden and Norway, *Evernia vulpina* is used for dyeing woollen stuffs yellow. Iceland Moss, *Cetraria Islandica*, is used in Iceland for dyeing brown. *Usnea barbata* is collected from trees in Pennsylvania, and used for an orange colour for yarn.

A general method for using lichens is suggested by Dr. Westring of Sweden in his *Experiments on Lichens for Dyeing Wools and Silks*:

"The Lichens should be gathered after some days of rain, they can then be more easily detached from the rocks. They should be well washed, dried, and reduced to a fine powder: 25 parts of pure river water are added to 1 of powdered lichen and 1 part of fresh quick lime to 10 parts powdered lichen. To 10 lbs. lichen half a pound sal ammoniac is sufficient when lime and sal ammoniac are used together. The vessel containing them should be kept covered for the first 2 or 3 days. Sometimes the addition of a little common salt or salt-petre will give greater lustre to the colours."

This method can be followed by anyone wishing to experiment with Lichens.

Dr. Westring did not use a mordant as a rule. Where the same species of Lichen grows on both rocks and trees, the specimens taken from rocks give the better colours.

Orchil or Archil and Cudbear are substantive or non mordants dyes, obtained from Lichens of various species of Roccella growing on rocks in the Canary Islands and other tropical and sub-tropical countries. They used to be made in certain parts of Great Britain from various lichens, but the manufacture of these has almost entirely disappeared. They have been known from early times as dyes. They give beautiful purples and reds, but the colour is not very fast. The dye is produced by the action of ammonia and oxygen upon the crushed Lichens or weeds as they are called. The early way of producing the colour was by treating the Lichen with stale urine and slaked lime and this method was followed in Scotland. Orchil is applied to wool by the simple process of boiling it in a neutral or slightly acid solution of the colouring matter. 3% Sulphuric acid is a useful combination. Sometimes alum and tartar are used. It dyes slowly and evenly. It is used as a bottom for Indigo on wool and also for compound shades on wool and silk. For cotton and linen dyeing it is not used. It is rarely used by itself as the colour is fugitive, but by [Pg 19] using a mordant of tin, the colour is made much more permanent.

Many of the British lichens produce colours by the same treatment as is used for producing Orchil. Large quantities were manufactured in Scotland from lichens gathered in the Shetlands and Western Highlands. This was called Cudbear. The Species used by the Scottish Cudbear makers were generally *Lecanora tartarea* and *Urceolaria calcarea*; but the following lichens also give the purple colour on treatment with ammonia:—*Evernia prunastri, Lecanora pallescens, Umbilicaria vellea, U. pustulata, Parmelia perlata.* Several others give colours of similar character, but of little commercial value. The manufacture of Archil and Cudbear from the various lichens is simple in principle. In all cases the plant is reduced to a pulp with water and ammonia, and the mass kept at a moderate heat and allowed to ferment, the process taking two or three weeks to complete.

## RECIPES FOR DYEING WITH LICHENS

*To dye Brown with Crotal.* For 6-1/4 lbs. (100 ozs.) of wool. Dye baths may be used of varying strengths of from 10 to 50 ozs. of Crotal. Raise the bath to the boil, and boil for an hour. A light tan shade is got by first dipping the wool in a strong solution of Crotal, a darker shade by boiling for half-an-hour, and a dark brown by boiling for two hours or so. It is better, however, to get the shade by altering the quantity of Crotal used. The addition of sufficient oil of vitriol or acetic acid to make the bath slightly acid will be an improvement (a very small quantity should be used).

*To dye red with Crotal.* Gather the lichen off the rocks—it is best in winter. Put layers of lichen and wool alternately in a pot, fill up with water and boil until you get the desired tint. Too much crotal will make the wool a dark red brown, but a very pretty terra cotta red can be got. No mordant is required.

*To dye Pink from a bright yellow Lichen (Parmelia parietina).* Mordant the wool with 3% of Bichromate of Potash, then boil with the lichen for 1 hour or more. [Pg 20]

*To dye Brown from Crotal.* Boil the wool with an equal quantity of lichen for 1 or 1-1/2 hours. No mordant is required.

*To dye red purple from Cudbear and Logwood.* Dye with equal quantities of Cudbear and Logwood, the wool having been mordanted with chrome. A lighter colour is got by dyeing with 8 lbs. cudbear, 1/2 lb. logwood (for 30 lbs. wool).

*To dye Yellow on Linen with the Lichen Peltigera canina* (a large flat lichen growing on rocks in woods). Mordant with alum (1/4 lb. to a lb. of linen) boil for 2 hours. Then boil up with sufficient quantity of the lichen till the desired colour is got.

## LIST OF LICHENS USED BY THE PEASANTRY OF DIFFERENT COUNTRIES FOR WOOL DYEING [D]

### SHADES OF RED, PURPLE AND ORANGE

*Borrera ashney.* Chutcheleera. India.

*Conicularia aculeata.* var. *spadicea.* Brown prickly cornicularia. Canary Islands, Highland Mountains.

*Evernia prunastri.* Ragged hoary Lichen. Stag's horn Lichen. Found in Scotland on trees.

*Isidium corallinum.* White crottle. On rocks in Scotland.

*I. Westringii.* Westring's Isidium. Norway and Sweden.

*Lecanora tartarea.* Crotal, Crottle, Corkur, Corcir, Korkir. Found in the Scotch Highlands and Islands, growing on rocks; used for the manufacture of Cudbear in Leith and Glasgow.

*L. parella.* Light Crottle, Crabs Eye Lichen. Found in Scotland, France and England, on rocks and trees; formerly celebrated in the South of France in the making of the dye called Orseille d'Auvergne.

*L. hæmatomma.* Bloody spotted lecanora, Black lecanora. Found in Scotland on rocks and trees.

*Lecidea sanguinaria.* Red fruited lecidea. In Scotland, on rocks.

[Pg 21]

*Nephroma parilis.* Chocolate coloured nephroma. Scotland, on stones. Said to dye blue.

*Parmelia caperata.* Stone Crottle, Arcel. Found in North of Ireland and Isle of Man, on trees. Said to dye brown, orange lemon and yellow.

*P. conspersa.* Sprinkled parmelia. Found growing on rocks in England.

*P. omphalodes.* Black Crottle, Cork, Corker, Crostil or Crostal (Scotch Highlands). Arcel (Ireland). Kenkerig (Wales). Alaforel leaf (Sweden). Found on rocks, especially Alpine, in Scotland, Ireland, Wales and Scandinavia. One of the most extensively used dye lichens. It yields a dark brown dye readily to boiling water, and it is easily fixed to yarns by simple mordants. It is stated to yield a red, crimson or purple dye.

*P. saxatilis.* Crottle, stane-raw, Staney-raw (Scotland). Scrottyie (Shetland). Sten-laf, Sten-mossa (Norway and Sweden). Found on rocks and stones in Scotland, Shetland, and Scandinavia. In winter the Swedish peasantry wear home made garments dyed purple by this Lichen. By the Shetlanders it is usually collected in August, when it is considered richest in colouring matter.

*Ramalina farinacea.* Mealy ramalina. On trees in England.

*R. scopulorum.* Ivory-like ramalina. Scotland, on maritime rocks. A red dye.

*Rocella tinctoria.* Orseille. Grows in the South of France, on the rocks by the sea.

*Solorina crocea.* Saffron yellow solorina. In Scotland, on mountain summits. The colouring matter is ready formed and abundant in the thallus.

*Sticta pulmonacea.* On trees.

*Umbilicaria pustulata.* Blistered umbilicaria. Found on rocks in Norway and Sweden.

*Urceolaria calcarea.* Corkir, Limestone Urceolaria. Found in Scotland, Western Islands, Shetland and Wales, growing on limestone rocks.

*U. cinerea.* Greyish Urceolaria. In England, on rocks.

[Pg 22]

*U. scruposa.* Rock Urceolaria. Grows on rocks in hilly districts in England.

*Usnea barbata.* Bearded Usnea. Pennsylvania and South America. On old trees. Stated to dye yarn orange.

*U. florida.* Flowering Usnea. Pale greenish yellow or reddish brown.

*U. plicata.* Plaited Usnea. On trees.

## SHADES OF BROWN

*Alectoria jubata.* Horsehair Lichen, Rock hair. On fir trees in England. Pale greenish brown.

*Borrera flavicans.* Yellow borrera. On trees in Germany. Gamboge yellow.

*Cetraria Islandica.* Iceland moss. Iceland heaths and hills. It yields a good brown to boiling water, but this dye appears only to have been made available in Iceland.

*Cetraria juniperina.* En-mossa. On trees in Scandinavia.

*Evernia flavicans.* Wolf's-bane evernia. On trees in Scandinavia. Gamboge yellow.

*Gyrophora cylindrica.* Cylindrical gyrophora. On rocks in Iceland. Greenish brown. Also G. deusta.

*G. deusta.* Scorched-looking gyrophora. On rocks in Scandinavia. Linnaeus states that it furnishes a paint called "Tousch", much used in Sweden.

*Lecanora candelaria.* Ljus mässa. On trees in Sweden.

*Lecidea atro-virens.* Map lichen. On rocks, Scandinavia.

*Lepraria chlorina.* Brimstone coloured lepraria. Scandinavia, on rocks.

*L. Iolithus.* Viol-mässa. Sweden, on stones. Gives stones the appearance of blood stains.

*Parmelia omphalodes.* In Scandinavia and Scotland. Withering asserts it yields a purple dye, paler, but more permanent, than orchil; which is prepared in Iceland by steeping in stale lye, adding a little salt and making it up into balls with lime.

*P. parietina.* Common yellow wall lichen, Wäg-mässla Wag-laf. England and Sweden, on trees, rocks, walls, palings. Used to dye Easter eggs. Used in Sweden for wool dyeing.

[Pg 23]

*P. physoides.* Dark crottle, Bjork-laf. Found in Sweden, Scotland and Scandinavia, on rocks and trees.

*Sticta pulmonacea.* Oak lung, Lungwort, Aikraw Hazelraw, Oak-rag, Hazel crottle, Rags. Found on trees in England, Scotland, North of Ireland, Scandinavia. It dyes wool orange and is said to have been used by the Herefordshire peasantry to dye stockings brown. Some species yield beautiful saffron or gamboge coloured dyes, e.g. *S. flava crocata, aurata.*

*S. scrobiculata.* Aik-raw, Oak rag. Found on trees in Scotland and England.

## FOOTNOTES:

[D] From an article by Dr. Lauder Lindsay on "The Dyeing Properties of Lichens." *The Edinburgh Philosophical Journal,* July to October, 1855.

[Pg 24]

# CHAPTER V

# BLUE

## INDIGO, WOAD, LOGWOOD

### *INDIGO*

Indigo is the blue matter extracted from a plant *Indigofera tinctoria* and other species, growing in Asia, South America and Egypt. It reaches the market in a fine powder, which is insoluble in water. There are two ways of dyeing with Indigo. It may be dissolved in sulphuric acid or oil of vitriol, thereby making an indigo extract. This process was discovered in 1740. It gives good blue colours but is not very permanent, darker colours are more so than the paler. It does not dye cotton or linen.

The other method is by the Indigo vat process which produces fast colours but is complicated and difficult. In order to colour with indigo it has to be deprived of its oxygen. The deoxidized indigo is yellow and in this state penetrates the woollen fibre; the more perfectly the indigo in a vat is deoxidized, the brighter and faster will be the colour. For wool dyeing the vats are heated to a temperature of 50°C. Cotton and linen are generally dyed cold.

### TO MAKE EXTRACT OF INDIGO

1 lb. oil of vitriol (pure, not commercial).
2 oz. finely ground Indigo.
1/2 oz. precipitated chalk.

Mix a little of the indigo with a small quantity of oil of vitriol, add a little chalk and stir well. Go on mixing gradually till all is used up. This should take an hour or two. Stir a few times each day for 4 or 5 days, adding about 1/2 oz. more of chalk by degrees. It is best mixed in a glass stoppered bottle or jar, and stirred with a glass rod. It must be kept from the air. [Pg 25]

**INDIGO EXTRACT (4 to 6 lbs. wool).**

Mordant [E] 25% Alum. Stir 2 to 3 ozs. Indigo extract into the water of dye bath. The amount is determined by the depth of shade required. When warm, enter the wool and bring slowly to boiling point (about 1/2 an hour) and continue boiling for another 1/2 hour. By keeping it below boiling point while dyeing, better colours are got, but it is apt to be uneven. Boiling levels the colour but makes the shade greener. This is corrected by adding to the dye bath a little logwood, 10 to 20 per cent which should be boiled up separately, strained, and put in bath before the wool is entered; too much logwood dims the colour. Instead of logwood a little madder is sometimes used; also Cudbear or Barwood.

**INDIGO VAT (TIN) FOR WOOL**

To 2 quarts of water add 1/4 lb. lime, and make hot. Then add 1 oz. indigo pounded up with a little of the lime water; let it stand and get warmer. Pound up 1/2 oz. tin, *Stannous Chloride*, in a little lime water and add, together with 1/2 oz. zinc. Add more lime water or tin according to the state of the vat. There should be a streaky scum on the surface, and the water underneath clear with a green tinge. Pearl ash can be used instead of lime.

**HYDROSULPHITE-SODA VAT FOR WOOL**

2 ozs. powdered indigo.
7 fluid ozs. Caustic Soda solution (SG 1.2).
4 pints Sodium Hydrosulphite (SG 1.1).

*The Stock Solution.*—Take 2 ozs. of well pounded indigo, with enough warm water (120°F.) to make a paste, and *grind* in a pestle and mortar for 10 minutes. Empty into a saucepan, capacity 1 gallon. Take 12 fluid ozs. of water adding gradually 3 ozs. of commercial caustic soda 76 per cent. This will give a solution of SG 1.2, which can be tested with a hydrometer reading from 1000 to 2000, the 1000 representing SG 1 as for water. [Pg 26]

Next take 5 pints water, add hydrosulphite slowly, stirring gently until a reading of 1100 is shown (SG 1.1) on the hydrometer. If the hydrosulphite be weighed beforehand and the stock of the same be kept free from damp air, or great heat, for future vats the hydrometer can be dispensed with; it is simply weighed out and added slowly to the water. If added too quickly the hydrosulphite will cake, fall to the bottom and be difficult to dissolve.

To the saucepan containing the indigo (100 per cent) add 7 fluid ozs. of the caustic soda solution, then gradually add 3-1/2 pints of hydrosulphite solution, stirring gently for 15 to 20 minutes. Heat the saucepan to 120°F. and on no account to more than 140°F. — *overheating will ruin the Stock Solution* — let it stand for half an hour, then test with a strip of glass. This should show a perfectly clear golden yellow colour (turning blue in 45 secs. approx.), free from spots. If dark spots show, this indicates undissolved indigo, therefore gradually add hydrosulphite solution (2-3 fluid ozs.). Wait 15 mins. and test with glass strip; if incorrect continue this every 15 minutes until the glass indicates clear yellow. If the Stock Solution is greenish white and turbid, undissolved *indigo white* is present. Add then not more than a teaspoonful at a time caustic soda solution until the Stock Solution answers the glass test.

The *Dye Vat* should contain about 10 gallons of water heated to hand hot, 120° and not above 140°F. Add 3 ozs. of hydrosulphite solution stirring carefully, let it stand for 20 minutes; this renders harmless any undissolved oxygen. Add a small cupful of the Stock Solution, stir carefully without splashing. The vats should be greenish yellow and *should not feel slimy*, an indication of too much caustic. The vat is now ready to dye and is kept at 120° to 140°F.

Between dips add Stock Solution as required, if the vat goes blue and turbid add 3 to 4 fluid ozs. of hydrosulphite and warm up to 140°F. and wait 30 minutes. As a last resort add caustic soda solution very gradually. This should not be required if the Stock Solution is properly prepared.

Start to dye with weak vats, 20 to 40 minute dips, and [Pg 27] finish with stronger vats. The more dips given to obtain a fixed shade, the faster will be the yarn to washing and rubbing. The yarn must be oxidized by exposure to the air for the same length of time as dipped. After the final dip, pass the yarn through a 10 gallon bath of water to which is added 3 ozs. of sulphuric acid, pure or hydrochloric. This neutralizes the caustic used. Wash yarn at least twice in water.



*Improvement of Defective Indigo Vat Dyes.*

If, after washing until clear, the yarn should rub off badly, there is but one remedy. Wash same in Fuller's earth, and if the shade is then too pale, re-dye. If, through bad management of the vats, the yarn is dull, pass the yarn through a hot bath (100% water, 1% acetic acid) and wash in two waters. If yarn is streaky, take 10 gallons of water at 120°F., 1 oz. of hydrosulphite powder, 2 fluid ozs. liquid ammonia fort. 880, and let yarn lie in same for 60 minutes. Wash in two waters.

The following facts should be carefully noted:—

The Caustic Soda is the *alkali* which dissolves the Indigo White.

The Hydrosulphite *reduces* the Indigotine in the Indigo to *Indigo White.*

Indigo White is fixed on the yarn as Indigo White and on exposure to the air becomes blue.

The yarn, on removal from the vat, should come out greenish yellow or a greenish blue. The latter is for blue yarn and should not turn blue too quickly (allow 60 seconds at least).

Rest the vats for 1 hour after 3 hours work. Never hurry the vats. It is a good thing to have hydrosulphite slightly in excess as this prevents premature oxidization; too much will strip off the indigo white already deposited on the yarn.

Caustic Soda must always be used with the greatest caution or the yarn will be tendered and ruined.

Finally, unless the yarn is completely scoured it is impossible to obtain a clear colour, or a blue which will not rub off. [Pg 28]

The figures given are for Indigo bearing 100% Indigotine, therefore in using vegetable Indigo do not add *all* the Caustic or Hydrosulphite, but depend on the glass test rather than on measurements.

## *WOAD*

Woad is derived from a plant, *Isatis tinctoria*, growing in the North of France and in England. It was the only blue dye in the West before Indigo was introduced from India. Since then woad has been little used except as a fermenting agent for the Indigo vat. It dyes woollen cloth a greenish colour which changes to a deep blue in the air. It is said to be inferior in colour to indigo but the colour is much more permanent. The leaves when cut are reduced to a paste, kept in heaps for about fifteen days to ferment, and then are formed into balls which are dried in the sun; these have a rather agreeable smell and are of a violet colour. These balls are subjected to a further fermentation of nine weeks before being used by the dyer. When woad is now used it is always in combination with indigo, to improve the colour. Even by itself, however, it yields a good and very permanent blue.

It is not now known how the ancients prepared the blue dye, but it has been stated (Dr. Plowright) that woad leaves when covered with boiling water, weighted down for half-an-hour, the water then poured off treated with caustic potash and subsequently with hydrochloric acid, yield a good indigo blue. If the time of infusion be increased, greens and browns are obtained. It is supposed that woad was "vitrum" the dye with which Caesar said almost all the Britons stained their bodies. It is said to grow near Tewkesbury, also Banbury. It was cultivated till quite lately in Lincolnshire. There were four farms in 1896; one at Parson Drove, near Wisbech, two farms at Holbeach, and one near Boston. Indigo has quite superseded it in commerce.

## LOGWOOD

### (*Bois de Campeche, Campeachy Wood*)

Logwood is a dye wood from Central America, used for [Pg 29] producing blues and purples on wool, black on cotton and wool, and black and violet on silk. It is called by old dyers one of the Lesser Dyes, because the colour was said to lose all its brightness when exposed to the air. But with proper mordants and with careful dyeing this dye can produce fast and good colours. Queen Elizabeth's government issued an enactment entirely forbidding the use of logwood. The person so offending was liable to imprisonment and the pillory. The principal use for logwood is in making blacks. The logwood chips should be put in a bag and boiled for 20 minutes to 1/2 an hour, just before using.

## RECIPES for DYEING with LOGWOOD

### (1). BLACK

Mordant the wool for 1 to 1-1/2 hours with 3 per cent Chrome and 1 per cent Sulphuric Acid. Wash and dye in separate bath for 1 to 1-1/2 hours with 50 per cent Logwood. This gives a blue black.

A dead black is got by adding 5 per cent Fustic to the dye bath.

A green black by adding more fustic. Also by adding 3 to 4 per cent Alum to the mordanting bath a still greener shade can be obtained.

A violet black is produced by adding 2 per cent Stannous Chloride to the dye bath and continue boiling for 20 minutes.

### (2). LAVENDER

Mordant with 3 per cent Bichromate of Potash for 45 minutes and wash. Dye with 2 per cent madder, 1 per cent logwood. Enter the wool, raise to the boil and boil for 45 minutes. The proportion of logwood to madder can be so adjusted as to give various shades of claret to purple.

### (3). A FAST LOGWOOD BLUE

(Highland recipe.) Mordant with 3 per cent Bichromate of Potash and boil wool in it for 1-1/2 hours. Wash and dry wool. Make a bath of 15 to 20 per cent logwood with about [Pg 30] 3 per cent chalk added to it. Boil the wool for 1 hour, wash and dry. The wool can be greened by steeping it all night in a hot solution of heather till the desired tint is obtained.

### (4). RAVEN GREY FOR WOOL

Mordant with 25 per cent Alum for 1/2 hour at boiling heat; then take it out, add to the same liquor 5 per cent copperas, and work it at boiling heat for 1/2 hour. Then wash. In another copper, boil 50 per cent logwood chips for 20 minutes. Put the wool into this for 1/2 hour; then return it into the alum and copperas for 10 to 15 minutes. Wash well.

### (5). DARK RED PURPLE WITH LOGWOOD

(2-1/2 lbs.) Mordant with 25 per cent alum and 1 per cent cream of tartar for 1 hour. Let cool in the mordant, then wring out and put away for 4 to 5 days.

Dye with 60 per cent logwood and 25 per cent madder. Boil up the logwood and madder in a separate bath and pour through a sieve into the dye bath. Enter the wool when warm and bring to the boil. Boil from 1/2 hour to 1-1/2 hours. Wash thoroughly in soft water.

### (6) PURPLE

(For 1 lb.) Mordant wool with 1/4 lb. alum and 1/2 oz. tartar for one hour; wring out and put away in a bag for some days. Dye with 1/4 lb. logwood for 1 hour.

## FOOTNOTES:

[E] If the Extract is used alone, a mordant is not essential.

[Pg 31]

# CHAPTER VI

# RED

## KERMES COCHINEAL MADDER

### *KERMES*

Kermes, or Kerms, from which is got the "Scarlet of Grain" of the old dyers, is one of the old insect dyes. It is considered by most dyers to be the first of the red dyes, being more permanent than cochineal and brighter than madder. In the 10th century it was in general use in Europe. The reds of the Gothic tapestries were dyed with it, and are very permanent, much more so than the reds of later tapestries, which were dyed with cochineal. Bancroft says "The Kermes red or scarlet, though less vivid, is more durable than that of cochineal. The fine blood-red seen at this time on old tapestries in different parts of Europe, unfaded, though many of them are two or three hundred years old, were all dyed from Kermes, with the aluminous basis, on woollen yarn."

Kermes consists of the dried bodies of a small scale insect, *Coccus ilicis*, found principally on the ilex oak, in the South of Europe, and still used there.

William Morris speaks of the "Al-kermes or coccus which produces with an ordinary aluminous mordant a central red, true vermilion, and with a good dose of acid a full scarlet, which is the scarlet of the Middle Ages, and was used till about the year 1656, when a Dutch chemist discovered the secret of getting a scarlet from cochineal by the use of tin, and so produced a cheaper, brighter and uglier scarlet."

Kermes is employed exactly like cochineal. It has a pleasant aromatic smell which it gives to the wool when dyed with it. [Pg 32]

### *COCHINEAL*

The dried red bodies of an insect (*Coccus Cacti*) found in Mexico are named Cochineal.

### (1). PURPLE, CRIMSON AND SCARLET

(For 1 lb. wool.) Mordant with Bichromate of Potash (3%). Dye for 1 to 2 hours with 3 oz. to 6 oz. cochineal. With alum mordant (25%) a crimson colour is got. With tin mordant (10%) a scarlet. With iron mordant (6%) a purplish slate or lilac.

### (2). SCARLET

Mordant with 6 per cent Stannous Chloride and 4 per cent Cream of Tartar, boiling 1 hour. Dye with 15 to 20 per cent Cochineal, boil for 1 hour.

Enter in both mordant and dye bath, cool, and raise slowly to the boil. To obtain a yellow shade of scarlet, a small quantity of Flavin, Fustic, or other yellow dye may be added to the dye bath.

### (3). SCARLET

(1 lb.) Into the same bath, put 1 oz. tin, 1/8 oz. oxalic acid, 4 oz. cochineal. Enter silk and boil for 1 hour. With less oxalic acid, a less scarlet colour will be obtained.

### (4). CRIMSON

Mordant with 20 per cent alum or with 15 per cent alum and 5 per cent Tartar. Dye in separate bath, after well washing, with 8 to 15 per cent cochineal. Boil 1 hour. A slight addition of ammonia to the dye bath renders the shade bluer.

### (5). ROSE RED

(1 lb.) Mordant with Alum. Dye with 2 oz. Madder, 2-1/2 ozs. Cochineal, 1/4 oz. Oxalic Acid and 1/2 oz. tin.

### (6). PURPLE (for 5 lbs.)

Mordant with 3 ozs. Chrome. Wash. Dye for 2 to 3 hours with 13 ozs. Cochineal, which has been boiled for 10 minutes before entering wool. A tablespoonful of vinegar added to the dye bath helps the colour. Wash thoroughly.

## *MADDER*

Madder consists of the ground-up dried roots of a plant *Rubia tinctorum*, cultivated in France, Holland and other [Pg 33] parts of Europe, as well as in India. Madder is one of the best and fastest dyes. It is used also in combination with other dyes to produce compound colours. The gradual raising of the temperature of the dye bath is essential in order to develop the full colouring power of madder; long boiling should be avoided, as it dulls the colour. If the water is deficient in lime, brighter shades are got by adding a little ground chalk to the dye bath, 1 to 2 per cent.

Madder is difficult to dye as it easily rubs off and the following points should be noted.

(1). The baths should be quite clean. Rusty baths must not be used.

(2). Before dyeing, the wool must be thoroughly washed so as to get rid of all superfluous mordant.

(3). A handful of bran to the pound of wool, helps to brighten the colour.

(4). The wool should be entered into a tepid dye bath and raised to boiling in 1 hour and boiled for 10 minutes or less.

### (1) RED

Mordant with 1/4 lb. Alum to the pound of wool. Boil for 1 hour, let cool in mordant, wring out and put away in bag for 3 or 4 days. Wash very thoroughly. Then dye with 5 to 8 ozs. madder according to depth of colour required, and a handful of bran for every pound of wool. Enter in cool bath and bring slowly to the boil in an hour or more. Boil for a few minutes.

### (2) ROSE RED

Mordant with Alum. Dye with 4 to 4-1/2 ozs. madder to lb. wool and a very small quantity of logwood (1/2 oz. to 1 oz. to 3 or 4 lbs. of wool).

## (3) BROWN

(1 lb.) Mordant with 2-1/2 ozs. Copper Sulphate. Dye with 2 ozs. to 4 ozs. Madder according to depth of colour required. For yellow brown add a small quantity of fustic (1/4 oz. to the lb.) [Pg 34]

## (4) RED BROWN

Mordant wool with 3% Chrome (see p. 9), wash well and dye with 5 to 8 ozs. madder, bringing slowly to the boil, and boil for 1 hour.

Various shades of brownish red can be got by a mixture of madder, fustic and logwood with a Chrome mordant in varying proportions such as 28 per cent Madder, 12 per cent Fustic, 1 per cent Logwood for a brownish claret. 5 per cent Madder, 4 per cent Fustic, 1/2 per cent Logwood for tan.

## *BRAZIL WOODS*

Various leguminous trees, including lima, sapan and peach wood, dye red with alum and tartar, and a purplish slate colour with bichromate of potash. Some old dyers use Brazil wood to heighten the red of madder.

*CAMWOOD, BARWOOD, SANDALWOOD, or SANDERSWOOD,* are chiefly used in wool dyeing, with other dye woods (such as Old Fustic, and logwood) for browns. They dye good but fugitive red with bichromate of potash, or alum.

[Pg 35]

# CHAPTER VII

# YELLOW

## WELD OLD FUSTIC TURMERIC QUERCITRON DYER'S BROOM HEATHER AND OTHER YELLOW DYES

Weld, *Reseda luteola*, is an annual plant growing in waste places. The whole plant is used for dyeing except the root. It is the best and fastest of the yellow natural dyes.

The plant is gathered in June and July, it is then carefully dried in the shade and tied up in bundles. When needed for dyeing it is broken into pieces or chopped finely, the roots being discarded, and a decoction is made by boiling it up in water for about 3/4 hour. It gives a bright yellow with alum and tartar as mordant. With chrome it yields an old gold shade; with tin it produces more orange coloured yellows; with copper and iron, olive shades. The quantity of weld used must be determined by the depth of colour required. Two per cent of stannous chloride added to the mordant gives brilliancy and fastness to the colour. Bright and fast orange yellows are got by mordanting with 8 per cent stannous chloride instead of alum. With 6 per cent copper sulphate and 8 per cent chalk, weld gives a good orange yellow. Wool mordanted with 4 per cent of ferrous sulphate and 10 per cent tartar and dyed in a separate bath with weld with 8 per cent chalk, takes a good olive yellow. 8 per cent of alum is often used for mordant for weld. A little chalk added to the dye bath makes the colour more intense; common salt makes the colour richer and deeper.

Weld is of greater antiquity than most, if not all, other natural yellow dyes. It is cultivated for dyeing in France, Germany and Italy. It is important as it dyes silk with a fast colour. [Pg 36]

## (1) OLD GOLD

Mordant with 2 per cent chrome and dye with 60 per cent of weld in a separate bath. 3 per cent chalk adds to intensity of colour.

### (2) YELLOW

Mordant with alum, and dye with 1 lb. of weld for every pound of wool. Common salt deepens the colour. If alum is added to the dye bath, the colour becomes paler and more lively. Sulphate of iron inclines it to brown.

### (3) ORANGE

Mordant with alum with a little weld in the bath. Dye with weld. Add teaspoonful of tin to the dye bath. Boil in separate bath with 1/4 oz. madder or cochineal to the pound.

## *OLD FUSTIC*

Fustic is the wood of *Morus tinctoria*, a tree of Central America. It is used principally for wool. With Bichromate of Potash as mordant, Old Fustic gives old gold colour. With alum it gives yellow, inclining to lemon yellow. The brightest yellows are got from it by mordanting with tin. With copper sulphate it yields olive colours (4 to 5 per cent copper sulphate and 3 to 4 per cent tartar). With ferrous sulphate darker olives are obtained (8 per cent ferrous sulphate). For silk it does not produce as bright yellows as weld, but can be used for various shades of green and olive. Prolonged dyeing should always be avoided, as the yellows are apt to become brownish and dull.

## RECIPES FOR DYEING WITH OLD FUSTIC

### (1) OLD GOLD

Boil the wool with 3 to 4 per cent chrome for 1 to 1-1/2 hours. Wash, and dye in a separate bath for 1 to 1-1/2 hours at 100°C. with 20 to 80 per cent of old fustic. [Pg 37]

### (2) OLD GOLD

Mordant with 3 per cent chrome, for 3/4 hour and wash. Dye with 24 per cent fustic and 4 per cent madder for 45 minutes.

## (3) BRIGHT YELLOW

Mordant wool with 8 per cent of stannous chloride for 1 to 1-1/2 hours, and 8 per cent of tartar. Wash, and dye with 20 to 40 per cent of fustic.

## (4) GREENISH YELLOW

Mordant wool with 3 per cent chrome, for 3/4 hour and wash. Dye with 6 per cent fustic, 33 per cent logwood. Boil 3/4 hour.

## (5) YELLOW

Mordant with 25 per cent alum, wash after laying by for 2 days, dye with 5 to 6 oz. fustic to lb.

## *TURMERIC*

Turmeric is a powder obtained from the ground-up tubers of *Curcuma tinctoria*, a plant found in India and other Eastern countries. It gives a brilliant orange yellow, but has little permanence. It is one of the substantive colours and does not need any mordant. Cotton has a strong attraction for it, and is simply dyed by working in a solution of Turmeric at 60°C. for about 1/2 hour. With silk and wool it gives a brighter colour if mordanted with alum or tin. Boiling should be avoided. It is used sometimes for deepening the colour of Fustic or Weld, but its use is not recommended, as although it gives very beautiful colours, it is a fugitive dye.

## *QUERCITRON*

Quercitron is the inner bark of the *Quercus Nigra* or Q. tinctoria, a species of oak growing in the United States and Central America. It was first introduced into England by Bancroft in 1775 as a cheap substitute for weld. He says, [Pg 38]

"The wool should be boiled for the space of 1 or 1-1/4 hours with one sixth or one eighth of its weight of alum; then, without being rinsed, it should be put into a dyeing vessel with clean water and also as many pounds of powdered bark (tied up in bag) as there were used of alum to prepare the wool, which is then to be turned in the boiling liquor until the colour appears to have taken sufficiently: and then about 1 lb. clean powdered chalk for every 100 lbs. of wool may be mixed with the dyeing liquor and the operation continued 8 or 10 minutes longer, when the yellow will have become both lighter and brighter by this addition of chalk."

Flavin is extract of Quercitron bark, and is much used for bright yellow with tin.

### YELLOW (1 lb.)

Mordant with alum. Dye with 1 oz. Flavin.

### ORANGE WITH FLAVIN OR QUERCITRON (1 lb.)

Put into bath first 1/2 oz. Cream of Tartar. Then 3/4 oz. tin mixed with water (important to enter the Tartar first). Enter yarn and boil for 45 minutes. In the meantime have mixed up 1/2 oz. Flavin and 1/2 oz. to 3/4 oz. Cochineal (according to depth of orange required) with 1/4 oz. tin with a little warm water. Remove yarn, enter flavin, madder and tin, take off the boil, enter yarn and stir well. Boil 30 minutes.

## *BARBERRY*

The roots and bark of *Berberis Vulgaris* is used principally for silk dyeing, without a mordant. The silk is worked at 50° to 60°C. in a solution of the dye wood slightly acidified with sulphuric, acetic or tartaric acid. For dark shades mordant with stannous chloride.

## *DYER'S BROOM*

*Genista Tinctoria.* The plant grows on waste ground. It should be picked in June or July and dried. It can be used with an alum and tartar mordant and gives a good bright yellow. It is called greening weed and used to be much used for greening blue wool. [Pg 39]

## *PRIVET*

*Ligustrum Vulgare.* The leaves dye a good fast yellow with alum and tartar.

## HEATHER

Most of the heathers make a yellow dye, but the one chiefly used is the Ling, *Calluna vulgaris*. The tips are gathered just before flowering. They are boiled in water for about half-an-hour. The wool, previously mordanted with alum or chrome according to the shade of yellow wanted, is put into the dye bath with the boiling liquor, which has been strained. It is then covered up closely and left till the morning. Or the wool can be boiled in the heather liquor till the desired colour is obtained.

## ONION SKINS

Prepare by mordanting with alum. Take a sufficient quantity of onion skins and boil for 30 minutes. This gives a good yellow. The addition of tin will make the colour more orange.

[Pg 40]

# CHAPTER VIII

# BROWN AND BLACK

## CATECHU ALDER BARK SUMACH WALNUT PEAT SOOT LOGWOOD AND OTHER DYES

### *CATECHU*

Catechu (Cutch) is an old Indian dye for cotton. It can also be used for wool and silk, and gives a fine rich brown. It is obtained from the wood of various species of Areca, Acacia and Mimosa trees. Bombay Catechu is considered best for dyeing purposes.

Catechu is soluble in boiling water. It is largely used by the cotton dyer for brown, olive, drab, grey and black. (See pp. 46, 47, 48.)

### LIGHT GREY

(For 6 lbs.) 1 oz. cutch, 1 oz. iron. Boil for 1/2 an hour in the cutch, then put into boiling iron, being very careful to stir well. Wash very thoroughly.

These proportions can be varied according to the shade of grey required; the more iron makes the colour browner, the more cutch the bluer grey.

### CATECHU BROWN

The wool is boiled for 1 to 1-1/2 hours, with 10 to 20 per cent catechu, then sadden with 2 to 4 per cent of copper sulphate, ferrous sulphate, or chrome, at 100°C., in a separate bath for 1/2 hour.

### *ALDER BARK*

The bark and twigs of alder are used for dyeing brown and black. For 1 lb. wool use 1 lb. alder bark. Boil the wool with it for 2 hours, when it should be a dull reddish brown. Add 1/2 oz. copper as for every pound of wool for black. [Pg 41]

## SUMACH

Sumach is the ground up leaves and twigs of the *Rhus coraria* growing in Southern Europe. It dyes wool a yellow and a yellow brown, but it is chiefly used in cotton dyeing.

## WALNUT

The green shell of the walnut fruit and the root are used for dyeing brown. The husks to be used for dyeing must be collected green and fresh, then covered with water and kept from the light to prevent them oxidizing. In the walnut tree there is an astringent colourless substance which gives a greenish yellow dye. This has the property of absorbing oxygen from the air and turning dark brown. It is only the unoxidized pale greenish stuff that can act as the dye, the dark brown itself has no affinity for the wool. Acids should be added to the dye bath to prevent oxidization. Without a mordant the colour is quite fast, but if the wool is mordanted with alum a brighter and richer colour is got. When used they are boiled in water for 1/4 hour, then the wool is entered and boiled till the colour is obtained. Long boiling is not good as it makes the wool harsh. It is much used as a "saddening" agent; that is, for darkening other colours.

"The best and most enduring blacks were done with this simple dye stuff, the goods being first dyed in the indigo or woad vat till they were a very dark blue, and then browned into black by means of the walnut root."—*William Morris.*

PEAT SOOT gives a good shade of brown to wool. Boil the wool for 1 to 2 hours with peat soot. Careful washing is required in several changes of water. It is used sometimes for producing a hazel colour, after the wool has been dyed with weld and madder.

OAK BARK. Mordant with alum and dye in a decoction of oak bark.

ONION SKINS. (Brown.) Mordant the wool with alum. Drying two or three times in between makes the colour more durable. Dry. Wash. Boil a quantity of onion [Pg 42] skins, and cool; then put in wool and boil lightly for 1/2 an hour to 1 hour; then keep warm for a while. Wring out and wash.

BLACK. Mordant with 3% Bichromate of Potash for 45 minutes. Dye with 1 oz. Hematin crystals, 3/4 oz. madder, 1/2 oz. Persian berries. After boiling for 1 hour remove wool and add 1/4 oz. cream of tartar, 1 oz. cochineal, 3/4 oz. iron, 1/2 oz. copper sulphate. Return wool and boil again for 1/2 hour. Wash in soap.

## VARIOUS RECIPES

MADDER for BROWN. (1 lb. wool.) Mordant with 1 oz. copperas and 1 oz. cream of tartar. Dye with 6 ozs. madder.

MADDER, etc., for FRENCH BROWN. Mordant with 3 per cent chrome. Dye with 8 per cent fustic, 2 per cent madder, 1 per cent cudbear, 2 per cent tartar. If not dark enough add 1 per cent logwood. Boil for 1/2 hour. Wash and dry.

TAN SHADE. (6-1/2 lbs. wool.) Mordant with 3 ozs. Chrome for 45 minutes and wash in cold water. Boil for 1/2 hour in a bag 5 oz. madder, 4 oz. Fustic, 1/2 oz. logwood. Enter the wool, raise to the boil, and boil for 45 minutes. By altering the proportions of madder and fustic various shades of brown can be got.

GREENISH BLACK. (For 1 lb.) Mordant with 3 per cent Chrome. Dye with 2 ozs. Fustic, 2 ozs. logwood, 1 oz. madder, and 1 oz. copperas.

DARK GREENISH-BROWN. (1 lb.) Mordant with 3 per cent chrome. Dye with 2 ozs. logwood, 4 ozs. madder, 1 oz. fustic, 1-1/2 ozs. copperas. Boil for 1 hour.

[Pg 43]

# CHAPTER IX

# GREEN

Green results from the mixing of blue and yellow in varying proportions according to the shade of colour required.

Every dyer has his particular yellow weed with which he greens his blue dyed stuff. But the best greens are undoubtedly got from weld and fustic.

The wool is first dyed in the blue vat; then washed and dried; then after mordanting, dyed in the yellow bath. This method is not arbitrary as some dyers consider a better green is got by dyeing it yellow before the blue. But the first method produces the fastest and brightest greens as the aluming after the blue vat clears the wool of the loose particles of indigo and seems to fix the colour.

If a bright yellow green is wanted, then mordant with alum after the indigo bath; if olive green, then mordant with chrome.

The wool can be dyed blue for green in three different ways:—1st in the Indigo vat, 2nd with Indigo Extract with Alum mordant, 3rd with logwood with Chrome mordant. For a good bright green, dye the wool a rather light blue, then wash and dry; Mordant with alum, green it with a good yellow dye, such as weld or fustic, varying the proportion of each according to the shade of green required. Heather tips, dyer's broom, dock roots, poplar leaves, saw wort are also good yellows for dyeing green. If Indigo Extract is used for the blue, fustic is the best yellow for greening, its colour is less affected by the sulphuric acid than other yellows.

According to *Bancroft*, Quercitron is the yellow above all others for dyeing greens. He says:— "The most beautiful Saxon greens may be produced very cheaply and expeditiously by combining the lively yellow which results from Quercitron bark, murio sulphate of tin and alum, with the [Pg 44] blue afforded by Indigo when dissolved in sulphuric acid, as for dyeing the Saxon Blue."

"For a full bodied green" he says "6 or 8 lbs. of powdered bark should be put into a dyeing vessel for every 100 lbs. wool, with a similar quantity of water: When it begins to boil, 6 lbs. murio-sulphate of tin should be added (with the usual precaution) and a few minutes afterwards 4 lbs. alum: these having boiled 5 or 6 minutes, cold water should be added, and then as much sulphate of Indigo as needed for the shade of green to be dyed, stirring thoroughly. The wool is then put into the liquor and stirred briskly for half an hour. It is best to keep the water just at the boiling point."

## RECIPES FOR DYEING GREEN

### (1) GREEN WITH QUERCITRON FOR WOOL

Dye the wool blue in the indigo vat, wash well. For 100 parts of wool put 3 of chalk and 10 or 12 of alum. Boil wool in this 1 hour. Then to same bath add 10 to 12 parts quercitron and continue boiling for 15 minutes, then add 1 part of chalk, this addition is repeated at intervals of 6 to 8 minutes till a fine green is brought out.

### (2) WITH INDIGO EXTRACT AND WELD FOR WOOL

Mordant 1 lb. wool with 4 ozs. alum and 1/2 oz. cream of tartar. Dye blue with sufficiency of indigo extract, wash and dry. Prepare a dye bath with weld which has been previously chopped up and boiled. Enter wool and boil for half an hour or more.

### (3) GREEN FOR WOOL

Mordant with alum and cream of tartar, add to the mordanting bath a little weld or fustic. Dye with 6 ozs. fustic (or weld). Dye in a separate bath with indigo extract, a rather bluer green than is wanted. Then put into a yellow bath till the right shade of green is got. [Pg 45]

### (4) GRASS GREEN

For 1 lb. wool: 1-1/2 oz. alum, 1/2 oz. sulphuric acid, 1/2 oz. salt, 1/4 oz. Tin crystals. Dissolve tin in separate saucepan and mix half of it with 1/4 oz. Flavin, add both to the bath together with indigo extract (1/2 tablespoonful). When hot enter yarn and boil hard for 1 to 1-1/2 hours. It turns a green when exposed to air. Wash very thoroughly.

## (5) JADE GREEN (1 lb.)

Mordant with 1/3 oz. Cream of Tartar and 4 oz. Alum for 1/2 hour. Take out wool and air. Cool bath a little and add half the amount of the indigo extract to be used (according to shade of green required, 1/2 oz. indigo extract makes a good colour). Enter wool and stir rapidly for 5 minutes or so without boiling. Take out wool. Mix in the rest of the indigo extract. Enter wool and boil for 10 minutes. Take out wool. Throw away a quarter of the water and add some with 3/4 oz. fustic extract. Enter wool and boil for 1/2 hour to an hour.

[Pg 46]

# CHAPTER X

# THE DYEING OF COTTON

The dyeing of cotton is difficult with the natural dye stuffs, there are only a few colours which can be said to be satisfactory. The fastest known in earlier days was Turkey red, a long and difficult process with madder and not very practical for the small dyer. It had its origin in India where it is still used; red Indian cotton is one of the fastest colours known. Catechu is another excellent cotton dye used for various shades of brown, grey and black. A cold indigo vat is used for blue, Indigo Extract is not used. Yellows can be got with weld, flavin, turmeric (for which cotton has a strong attraction), and fustic. Great care is to be taken in dyeing yellow as it is not very fast to light. Greens may be got by dyeing in the indigo vat and then with a yellow recipe, purples from logwood with tin mordant, but purples and greens are unsatisfactory, and not suitable to the vegetable dyer.

## BOILING OUT

Before dyeing cotton in the raw state, or in yarn spun direct from the raw state, it must be boiled for several hours to extract its natural impurities. For dark colours water alone may be used, but for light and bright colours a weak solution of carbonate of soda, 5%; or of caustic soda, 2%, should be used.

## MORDANTS

*Alum.* Alum (1/4 weight of cotton) is dissolved in hot water with carbonate of soda crystals, or other alkali (1/4 weight of alum); work cotton in the solution, steep for several hours or overnight. Then well wash. Aluminium acetate solution as for silk (page 56) may be used. After drying, the cotton may be passed through a fixing solution of some alkali, for examples [Pg 47] see page 50. Before mordanting with alum, the cotton is often prepared with tannic acid.

*Iron.* Iron is usually employed as a "saddening" agent, i.e. the cotton after dyeing is steeped in a cold solution of the mordant. A further use is in dyeing black, when the cotton, after being prepared with tannin, is steeped in a cold solution of Iron. This process by itself gives a dark colour before any dye is used.

*Tin.* Tin is rarely used alone as a mordant for cotton but brightens the colour in combination with other mordants.

*Chrome.* Chrome is used for browns and other colours with Catechu. After boiling in a solution of the dye stuff, boil a short time in chrome solution, this oxidizes the colouring matter of the Catechu.

*Copper.* Copper is sometimes added in small quantities to the dye bath for brown or yellow to vary the shade.

*Tannin (Tannic acid).* Cotton and linen strongly attract tannin and when prepared with it they are able to retain dyes permanently. Cotton saturated with tannin attracts the dye stuff more rapidly, and holds it. Tannic acid is the best tannin for mordanting as it is the purest and is free from any other colouring matter; it is, therefore, used for pale and bright shades. But for dark shades, substances containing tannic acid are used, such as *sumach, myrobalans, valonia, divi-divi, oak galls, chestnut* (8 to 10 per cent tannin), *catechu.*

Cotton and linen are prepared with tannin after they have been through the required cleansing, and, if necessary, bleaching operations. A bath is prepared with 2 to 5 per cent of tannic acid of the weight of the cotton, and a sufficient quantity of water. For dark shades, 5 to 10 per cent should be used. The bath is used either hot or cold. It should not be above 60°C. The cotton is worked in this for some time, and then left to soak for 3 to 12 hours, while the bath cools. It is then wrung out and slightly washed.

The following gives the relative proportions of the various substances containing tannin: — 1 lb. tannic acid *equals* 4 lbs. sumach, 18 lbs. myrobalans, 14 lbs. divi-divi, 11 lbs. oak galls. [Pg 48]

*Examples from various recipes*:

For 10 lbs. cotton use 12 ozs. tannic acid.
" 50 " " " 10 lbs. sumach.
" 40 " " " 10 lbs. "

" 20 "   "   " 2 lbs. yellow (or black) catechu.
" 20 "   "   " 3 lbs. catechu with 3 ozs. blue vitriol.


Some recipes soak the cotton 24 hours, others 48.

## RECIPES FOR DYEING

### (1) INDIGO VAT

Take 3 oz. well ground indigo, mix into a paste with hot water. Slake 3 oz. Quicklime and boil with 6 oz. Potash or Soda ash in sufficient water, let it settle, pour off the clear liquor in which dissolve the indigo paste, boil or keep hot 24 hours; it should then have the consistency of thick cream, with much froth. During the boiling, slake another 3 oz. quicklime, boil in a pint of water for 15 minutes, let settle, pour off the clear liquor in which dissolve 4 to 5 oz. green copperas. Add the indigo and copperas solutions to 5 gallons water, stir well, let vat rest, stir once or twice during 24 hours or until it appear ready for dyeing. Before use it should be stirred and let stand 2 hours. It should be a clear yellowish green with much scum.

The cotton to be dyed should be entered in dips of increasing lengths of time, as 1, 5, 10, 20 minutes, and aired in between, according to depth of shade required. It should then be well washed, passing through water slightly acidulated with Sulphuric acid (a teaspoonful to 1 gallon). When this vat appears exhausted and turns a dark colour it may be revived by adding 2 or 3 oz. Green Copperas dissolved as before. When again exhausted, more of all the ingredients must be added.

### (2) LIME COPPERAS VAT

2 oz. Indigo, 4 oz. Copperas, 5 oz. Quicklime (fresh). Mix Indigo into a paste with hot water. Dissolve copperas in [Pg 49] hot water. Slake lime. Fill earthenware jar with about 5 gallons cold water and add the Indigo, copperas and slaked lime in that order. Stir well, cover and let stand till next day or until vat is in proper condition; it should be clear brownish yellow with possible blue scum. There will be some sediment. The dyeing process is as in (1).

## (3) RED

(For 1 lb. cotton.) The Turkey Red process is long and difficult. (1) Boil yarn 6 to 8 hours in a solution of carbonate of soda, 1-1/2 oz., wash well and dry. (2) Prepare a solution of 2 fluid ozs. Turkey Red oil, 2 ozs. ozrbonate of soda at 100°F., work cotton in this till thoroughly saturated, wring out, dry. (3) Repeat No. 2. (4) Repeat No. 2. (5) Steep 3 or 4 hours in solution of 1 oz. carbonate of soda at 100°F., wring out, dry. (6) Repeat No. 5 with a slight increase of soda. (7) as No. 6. (8) Steep 10 hours in water at 100°F., dry. The cotton should now be clear white. (9) Steep 4 hours in solution of 1-1/2 oz. tannic acid or 4 oz. Galls, at 100°F., wring out, dry. (10) Steep 24 hours in solution made by dissolving 10 oz. alum in hot water, and slowly adding 2-1/2 oz. carbonate of soda crystals, wring out and dry. The cotton is now grey coloured. (11) Dye with 2 lbs. madder. Bring slowly to the boil, boil for 1 hour, a white scum on the surface denotes the cotton has absorbed all its colour. A teaspoonful of chalk may be added to the dye-bath. The cotton is now dark claret colour. (12) To brighten, boil 3 or 4 hours in a solution of 1/2 oz. carbonate of soda crystals and 1/2 oz. soap. The bath should be covered, except for a small outlet for the steam which otherwise should be retained as much as possible. (13) The cotton can be further brightened by boiling with 1/2 oz. soap and a teaspoonful of Tin. Wash and dry.

## (4) RED

(For 1 lb.) After boiling out in soda, wash and dry. Steep overnight in a hot bath of 1-1/2 oz. Tannic acid or 4 oz. Galls, dry, steep in cold solution of 1/4 lb. alum and 1/2 oz. chalk, dry, add 2 oz. more alum to solution and steep as before, wash and dry. Dry with 12 oz. Madder, bring to boil in 1 hour and [Pg 50] boil a few minutes, rinse, re-dye as above, pass through warm soap bath, 2 oz., wash and dry.

## (5) YELLOW

(For 1 lb.) Mordant twice in Aluminium acetate, as described for silk (page 73), or in 1/4 lb. alum and 1-1/2 oz. chalk, steeping in cold solution. Pass through weak bath of chloride of lime, wash, dry. Dye with 2-1/2 lbs. weld and 1/2 oz. copper sulphate, boil for

1 hour, then boil with soap. Or dye with 2 to 3 oz. Quercitron, which should be brought slowly to the boil and boiled for a few minutes only.

## (6) YELLOW

(For 1 lb.) Steep overnight in hot bath of 1-1/2 oz. Tannic acid, or 4 oz. Galls, wring out, dry. Work 2 hours in bath of 1/4 lb. alum and 1/2 oz. chalk, dry, pass through weak bath of chloride of lime about 1 oz., dry. Return to alum bath and repeat process, wash well, dye slowly with 1-1/2 oz. Flavin.

## (7) ORANGE

(For 1 lb.) Boil 2 oz. Annatto with 1 oz. carbonate of soda crystals for 1/2 hour, then add to a bath containing a teaspoonful of Turkey Red Oil, boil for 10 minutes. Take off boil, enter yarn, boil for 1-1/4 hours, let cool to hand heat, remove yarn, wash slightly and dry quickly.

## (8) BROWN

(For 1 lb.) Enter in one bath 1 oz. Cutch, in another 1/2 oz. Chrome. Enter cotton in cutch bath, boil 20 minutes, wring out, boil 10 minutes in chrome bath. Add 6 oz. fustic or 1 oz. flavin to cutch bath, re-enter cotton. Repeat above until the required depth of colour is reached, finish in cutch bath to obtain deepest shade, which may be darkened by adding 1 drachm or so copper sulphate. A greyish drab may be got by adding ferrous sulphate. All shades of brown may be obtained by decreasing or increasing the amount of cutch or by adding a little logwood or fustic, in which latter case the cotton should have been previously mordanted. [Pg 51]

## (9) BLACK

(For 1 lb.) Wash, steep overnight in hot solution of tannic acid, 1 oz., wring out without washing, work for 10 minutes in soda bath, at a temperature of 50° to 60°C., 1-1/4 oz. Wring out, work in cold solution of copperas, 1-1/4 oz., for 1/2 hour, return to soda bath for 1/4 hour. Wash, dye in bath of logwood 12 oz., madder 2-1/2 oz., and fustic 8 oz. Enter into cold bath and raise gradually to boiling, boil for 1/2 hour, pass through warm solution of chrome, 1 oz., wash, work through warm soap bath.

Greys may be obtained with 1 to 5 per cent of logwood after mordanting in a weak solution of iron.

## THE ZINC-LIME INDIGO VAT

*The Zinc-lime Indigo Vat.* It will be necessary to explain these words—Indigo blue is insoluble and cannot be used for dyeing. If however it is "reduced" or changed to indigo white, it has, while it is in this form, an affinity for vegetable and animal fibre. These fibres will take it up from the solution and retain it. If they are then exposed to the air, the oxygen acts upon the indigo in the fibre and turns it back again to indigo blue. Various chemicals can be used to reduce indigo blue to indigo white. I propose to describe how the work is done with zinc dust and lime as reducing agents.

In course of time the word "vat" has been transferred from the dyeing vessels themselves to their contents; *i.e.,* the indigo dye liquor. By "vat," therefore, we understand not only the vessel used for dyeing indigo, but the solution of alkali salts of indigo white in water. This definition distinguishes the *indigo vat* completely from indigo extract, or any other improper purposes to which indigo may be put.

The zinc lime indigo vat is better than any other for dyeing cotton and linen. It is also very good for dyeing silk. It has many advantages over the hydrosulphite vat, as it is not nearly so much affected by changes of temperature and weather. It can be put to work after a six months' rest.

The disadvantage which it shares with the copperas vat, though in a less degree, is that there is a sediment which [Pg 52] must not touch the stuff during the dyeing. This is avoided by hanging a net in the vat after the sediment has settled, or by dipping the skeins on rods.

It is essential that the indigo used should be of the best quality, and ground to so fine a powder that it will float on water. Coarsely ground indigo will never reduce and can be found at the bottom of the vat unchanged. It should be so fine that no roughness is felt with the tongue. Buy the best quality indigo ready ground, and if possible mixed to a paste with water. A 20% paste, *i.e.* 20% of indigo and 80% of water, is a usual quantity. If indigo powder must be used it must be mixed to a paste very carefully, as it will, if properly ground, fly about like dust. The easiest method of mixing is to pour the required amount of boiling water into a jar (previously heated), then put in the indigo. Close the vessel tightly. The steam which rises will moisten the indigo so that it loses its tendency to fly about. After 10 or 15 minutes it can easily be mixed with a stick. The zinc dust should be dry and not caked.

*The lime* should be in hard lumps. It should be bought from a reliable chemist in a sealed container, and kept sealed till wanted. If it is crumbling and cracking it has been exposed to damp air, and is partly slaked already, and therefore more or less useless.

As the indigo is more quickly reduced in a concentrated solution, a stock vat is first made and this is added to the dye vat as required. The vessel for the stock vat should have a well-fitting lid. A stoneware jar with a bung will do very well. To make a stock vat sufficient to furnish a dye vat containing 15-20 gallons use: —

10 oz. Indigo 20% paste (or 2-1/2 oz. indigo pasted with
7-1/2 oz. of water),
1-1/2 oz. zinc dust,
4-5 oz. quick lime,
4-5 pints of water.


Mix the zinc dust to a paste with a little of the water, gradually add the indigo and the rest of the water. The heat of the water should be not less than 160°F. as it will cool while the [Pg 53] lime is being prepared. Slake the lime in a separate vessel by pouring about 5 oz. of water over it. When it begins to hiss and break, add more water little by little. When all the lumps have cracked up stir till a thick even cream is made. Add this to the other ingredients in the stock vat. Stir well. The stock vat should have a temperature of 120-140°F. It should be stirred at intervals. The vessel should be stood in hot water to keep the temperature as near 120°F. as possible. In about 5 hours the mixture has a pure yellow colour and is ready to add to the dye vat. (There is of course a blue-black scum of indigo on top.)

*Preparation of the dye vat.* The vessel used should be deep and upright so that an unnecessarily large surface is not exposed to the air, and a sufficient space for dyeing is obtained above the sediment. A galvanised dust bin, or a barrel (provided it is not of oak or any other wood which contains tannin), make good indigo vats. Put 16 gallons of water in the vat at a temperature of 65-70°F. In order to counteract the effects of the atmospheric oxygen contained in the water of the vat, additions of zinc dust and lime are made some hours before the stock solution is added. A pinch of zinc dust and an ounce of lime, previously slaked, should be added and the vat stirred. Stirring must always be done gently and smoothly, every effort being made not to take air into the vat. At the same time it must be stirred up from the bottom so that the sediment is mixed with the liquor above it. The best tool for this purpose is a broom stick, to one end of which a piece of wood is nailed, like a garden rake. When all is ready, carry the stock solution to the dye vat, and, to avoid splashing through the air, hold it in the water of the vat while gently pouring out half its contents. Stir up the vat and cover it until it shows a clear yellow colour under the surface of the scum. This may not happen for 24 hours. A good way to test the colour of the vat is to push back the scum with the edge of a saucer or plate, then dip it halfway into the liquor. Against its white surface the colour of the liquor will be plainly seen. It should look like good light ale. If the liquor is greenish and sufficient time has elapsed, [Pg 54] another pinch of zinc dust and a little more lime must be added as before, and the vat again stirred, allowed to settle and again tested. A little difficulty may be found in getting the vat to start, but once it has worked well no difficulty will be found in starting it again. It will work more easily as it gets older.

As indigo does not penetrate easily, every effort must be made to help it to do so. The stuff to be dyed must be thoroughly scoured so that no particle of grease, size, or any other impurity is present. Every effort must be made to prevent unreduced indigo from attaching itself to the cotton. Never begin to dye in a vat which is greenish. The unreduced indigo will attach itself to the stuff and be wasted. Your time will also be wasted in washing it off.

The vat should be thoroughly stirred and allowed to settle each day before dyeing begins. When the sediment has settled, the froth should be carefully skimmed and kept to return to the vat when the day's dyeing is finished.

If a net is to be used it should be thoroughly wetted (if everything goes into the vat wet it will take less air with it). The net can be kept down by tying a few stones in a bag or an iron weight to the centre of it. If the hanks are to be dipped on a rod this may be of iron, or of wood suitably weighted. The hanks should not be less than 8 inches below the surface of the liquor and about 1 ft. above the bottom of the vat. The hanks should be turned after each dip, as, if the same end goes to the bottom each time it will be darker. A pulley over the vat to draw out the rod or net is convenient. The dyeings can then be allowed to drain a few seconds. Then wring each hank, shaking it out to get the air into it. After a sufficient airing, dip again. Many short dips with airing between will produce faster colours. Dip 1 minute, wring and air 2 minutes. Dip 2 minutes, wring and air 4 minutes. Dip 5 minutes, and so on.

As linen and cotton look so very much darker when wet than when dry, a bit should be dried to judge if the colour is right. [Pg 55]

Indigo can be dyed from the palest sky blue to black. The very palest shade of sky blue is never very fast. The virtue which indigo alone seems to possess is that, though it may become lighter with continual use, it also becomes a clearer and more lovely blue. This is especially so on cotton and linen, for which it is a superb dye. The varying shades of indigo of butchers' coats, sailors' collars, and French porters' blouses always give us pleasure.

[Pg 56]

# CHAPTER XI

# THE DYEING OF SILK

Silk is covered with a natural gum which has to be removed before the dyeing process can begin. This is done by boiling for one hour or more in a bath containing soap, 2 to 8 ozs. to the pound of silk according to the amount of gum on the silk. It is then well washed, and is ready for mordanting.

The mordants mostly used are *Alum,* for most of the bright colours. *Tin,* for brightening some colours, and as a separate mordant for others. *Iron,* for black dyeing. *Chrome,* for certain browns such as catechu.

The principal Alum mordant is Acetate of Alumine, prepared as follows: Let 3 lbs. Alum and 3 ozs. chalk be dissolved in 1 gallon of warm water in an earthenware pan, add the chalk slowly to the Alum. Add 2 lbs. white acetate of lead, stir occasionally during 24 to 36 hours. Let it remain 12 hours at rest. Decant and preserve the clear liquor, being careful not to stir up the sediment. Pour 2 gallons of water on the sediment, and stir occasionally for 12 hours. Let it rest 12 hours. Decant the clear and add to the first lot. Bottle for use. It keeps about three weeks. Of the mordant 2 parts are diluted with 1 of water, and the silk is well worked in this for 10 minutes, after being wetted down. Steep for 12 hours, wring out and dry. Wet down again and return to the Alum liquor, work for 10 minutes, steep 12 hours, dry. When thoroughly dry, wash well in several changes of water before dyeing. For less bright colours one mordanting may be sufficient.

The mordant is used for successive batches of silk until exhausted; the fresher the mordant, the better for brighter colours. Silk should be dyed as soon after it is dried as is convenient.

[Pg 57]

Another Alum mordant. Dissolve 25 per cent of Alum in hot water and add 6 per cent carbonate of soda crystals. Fill up a jar with water and steep silk in it over-night. It must be washed before dyeing.

## RECIPES FOR DYEING

### (1) INDIGO VAT FOR BLUE

Silk is dyed in a similar manner as described for wool, but requires stronger vats and longer dips to obtain the same depth of colour. See page 33.

### (2) INDIGO EXTRACT FOR BLUE

Dye at a temperature of 40 to 50°C. with as much Indigo Extract dissolved in the bath as is required for the desired depth of shade. If the silk has been first mordanted with alum, compound colours can be obtained by the addition of a red or yellow dye to the bath.

### (3) CRIMSON

Mordant with Alum or Aluminium Acetate and dye with 40 to 50 per cent Cochineal. A teaspoonful of Tin, dissolved in cold water, may be added to brighten. Boil well. It is advisable to wash in soap after using tin as it prevents the latter making the yarn brittle.

### (4) MADDER RED

Mordant with Alum or Aluminium Acetate. Dye with 80 to 100 per cent Madder and a handful of bran per pound of silk. Bring slowly to the boil in 1 hour, boil a few minutes. It should be brightened by boiling a short time in soap, with a little tin.

### (5) YELLOW

Mordant with Alum or Aluminium Acetate. Various Dyes may be used. *Weld*: Dye with 150 per cent. *Flavin*: Dye with 1 oz. to the pound, with a teaspoonful Tin. *Fustic*: Dye with 50 per cent, or more. *Quercitron*: Dye with 10 to 20 per cent. A little chalk may be added towards the end. [Pg 58]

The shades may be varied by the addition of small quantities of madder or cochineal. Orange may be obtained by the use of Madder, 2 to 4 ozs. per pound, with Flavin or Fustic.

## (6) GREEN

Greens may be obtained by dyeing with any of the yellow dyes and blueing in the Indigo Vat or with Indigo Extract. If the colour is thin, it should be dyed a deeper blue in the vat and then re-dyed with yellow. A strong clear yellow is needed for a good green.

## (7) PURPLE

Dye silk blue in Indigo Vat. Then dye without mordanting in Cudbear.

## (8) ORANGE (1 lb.)

Mordant with Alum Acetate. Dye with 1/2 lb. Madder, 2 ozs. Flavin and 1 oz. tin.

Enter the tin first in a cold bath. Mix Flavin and Madder into a paste and add to the bath. Bring to the boil slowly, boil for 10 minutes. Wash in soap.

## (9) BLACK (1 lb.)

Mordant with Alum Acetate. Dye with 6 ozs. logwood, 3/4 oz. flavin, 1 oz. Iron. Mix all together and boil for 1/2 hour. Wash thoroughly.

## (10) BLACK

(1) Mordant with basic ferric sulphate and after allowing the silk to lie for some time, wash well and soap at 90°C.

(2) Dye with 50 per cent Fustic, 10 per cent Ferrous Sulphate and 2 per cent Copper Acetate.

(3) Dye with logwood 50 per cent and soap.

## (11) GREY WITH BRACKEN (1 lb.)

Mordant with 1 oz. Iron and 2 ozs. Cream of Tartar. Boil a quantity of young bracken tips for 1/2 hour. Strain. Boil silk in the decoction for about an hour. [Pg 59]

## (12) BROWN WITH LICHEN (1 lb.)

Mordant with Alum Acetate. Put into the dye bath the quantity of lichen according to required colour with about a teaspoonful of Acetic Acid. Boil from 1 to 3 hours.

## (13) ORANGE (1 lb.)

1 oz. tin, 1/2 oz. Oxalic Acid, 2 oz. Flavin. Enter silk and boil for 1 hour. Remove silk and add to the bath 1 oz. tin, 1 oz. Oxalic, 2 oz. Cochineal. Boil for 1 hour or more.

## (14) BLACK (1 lb.)

Mordant with 2 oz. logwood extract, 1-1/4 oz. fustic extract, 1-1/4 oz. iron, 1/2 oz. copper sulphate. Boil for 1 hour. Take out and rinse. To the same bath add 1-3/4 oz. logwood extract, 1 oz. fustic extract, 7 oz. madder. Enter silk and boil for 1 hour. Wash in soap.

## (15) YELLOW (1 lb.)

Mordant with 1 oz. Bichromate of Potash. Boil 1 hour. In a separate bath put 1 lb. weld and boil for 1 hour.

## (16) RED (1 lb.)

Mordant with 1-3/4 oz. tin and 1-3/4 oz. oxalic acid. Boil for 1 hour. Then add 3/4 lb. cochineal and 6 oz. madder. Boil well and wash in soap.

## (17) BROWN (1 lb.)

Mordant with 1 oz. Copper sulphate. Boil for 1 hour. Take out silk and add 2-1/2 oz. madder, 1 oz. fustic chips, and boil for 1 hour.

## (18) RED (1 lb.)

Dissolve 1 oz. Tannic Acid in hot water. Enter silk and leave for 24 hours, stirring occasionally. Rinse well in two waters. In a fresh bath, put 4 oz. cochineal. Enter silk. Bring to boil and let blue colour develop. Lift, and add 1 oz. cochineal & 1 oz. tin. Re-enter silk & boil well. Wash in soap.

[Pg 60]

# GLOSSARY

*Adjective Dyes.* Dyes which require mordant.

*Alizarin.* The chief colouring principle of madder. It is also the name for an extensive series of chemical colours produced from anthracene, one of the coal tar hydrocarbons discovered in 1868.

*Aniline.* Discovered 1826 (*anil. Span. indigo*). First prepared from indigo by means of caustic potash, found in coal, 1834. Manufactured on a large scale after Perkin's discovery of mauve in 1856.

*Annatta.* (Annotto, Arnotto, Roucou.) A dye obtained from the pulp surrounding the seeds of the *Bixa orellana*; chiefly used in dyeing silk an orange colour, but is of a fugitive nature.

*Argol.* The tartar deposited from wines completely fermented, and adhering to the sides of casks as a hard crust. When purified it becomes Cream of Tartar.

*Beck.* A large vessel or tub used in dyeing.

*Bois jaune.* Fustic, yellow wood.

*Carthamus.* Safflower, an annual plant cultivated in South Europe, Egypt and Asia, for the red dye from its flowers.

*Caustic Soda.* Carbonate of soda, boiled with lime.

*Coal Tar Colours.* Colours obtained by distillation and chemical treatment from coal tar, a product of coal during the making of gas. There are over 2,000 colours in use.

*Detergent.* A cleansing agent.

*Dip.* Generally applied to immersing cloth, etc., in the blue vat.

*Divi-Divi.* The dried pods of *Caesalpina coriaria* growing in the West Indies and S. America; they contain 20 to 35% tannin and a brown colouring matter.

*Dyer's Spirit.* Aqua fortis, 10 parts; sal ammoniac, 5 parts; tin, 2 parts; dissolved together.

*Enter.* To enter wool, to put it into the dye or mordant liquor.

*Fenugrec.* Fenugreek *Trigonnella fænugræcum.*

*Flavin.* A colouring matter extracted from Quercitron. [Pg 61]

*Full, to.* To treat or beat cloth for the purpose of cleansing and thickening it.

*Fuller's herb. Saponaria officinalis.* A plant used in the process of fulling.

*Fuller's Thistle*, or teasle. *Dipsacus fullonum.* Used for fulling cloth.

*Fustet.* Young fustic. Venetian Sumach *Rhus cotinus.* It gives a fine orange colour, which has not much permanence.

*Galls, Gall nuts.* Oak galls produced by the egg of an insect, — the female gall wasp. An excrescence is produced round the egg, and the insect, when developed, pierces a hole and escapes. Those gall nuts which are not pierced contain most tannic acid. The best come from Aleppo and Turkey.

*Gramme* or *Gram.* About 15-1/2 grains (Troy).

*Kilo. Kilogramme.* Equals 2 lbs. 3.2 oz.

*Litre.* Nearly 1-3/4 pints.

*Lixivitation.* The process of separating a soluble substance from an insoluble by the percolation of water.

*Lixivium.* (Lye.) A term often used in old dye books, water impregnated with alkaline salts extracted by lixivitation from wood ashes.

*Lye* or *Ley.* Any strong alkaline solution, especially one used for the purpose of washing such as soda lye, soap lye.

*Mercerised Cotton.* Cotton prepared by treating with a solution of caustic potash or soda or certain other chemicals. Discovered by John Mercer in 1844.

*Milling.* The operation of fulling cloth.

*Myrobalans.* The fruit of several species of trees, growing in China and the East Indies, containing tannic acid (25-40% tannin).

*Oil of Vitriol.* Sulphuric acid.

*Organzine.* Twisted raw silk from best cocoons, used for warp.

*Pearl Ash.* Carbonate of potash.

*Persian Berries.* The dried unripe fruit of various species of Rhamnus. Also called French berries, grains of Avignon. [Pg 62]

*Potassium Carbonate.* (Potashes.) Carbonate of potash has been known since ancient times as a constituent of the ashes of land plants, from which it is obtained by extraction with water. In most cases Sodium Carbonate, which it strongly resembles, can be used in its place.

*Red spirits.* Tin Spirits. Applied to tin mordants generally. A solution of Stannous chloride.

*Red woods.* Camwood, Barwood, Sanderswood (Santal, Sandal, Red Sanders), Brazil wood, Sapan wood, Peach-wood.

*Roucou.* Anatta, Arnotto.

*Saxon blue.* The dye made by indigo dissolved in oil of vitriol.

*Scotch ell.* 37.2 inches.

*Scour, to.* To wash.

*Scroop.* The rustling property of silk.

*Soda ash.* Carbonate of soda.

*Sour water.* To every gallon of water, add one gill vitriol; stir thoroughly. Stuff steeped in this should be covered with the liquor, otherwise it will rot.

(2) Water in which bran has been made to grow sour. 24 bushels of bran are put in a tub, about 10 hogsheads of nearly boiling water is poured into it; acid fermentation soon begins, and in 25 hours it is ready to use.

(3) Throw some handfuls of bran into hot water and let it stand for 24 hours, or until the water becomes sour, when it is fit for use.

*Staple.* A term applied to cotton and wool indicating length of fibre.

*Substantive Dye.* A dye not requiring a mordant.

*Sumach.* Leaves and twigs of several species of Rhus, containing tannic acid. It is sold in the form of crushed leaves or as a powder (15-20% tannin).

*Tram.* Slightly twisted raw silk, used for weft.

*Tyrian purple.* A purple colour obtained from certain shell fish, such as Buccinum and Purpura. It is mentioned by Pliny as being discovered in 1400 b.c. It was a lost art in the Middle Ages. [Pg 63]

*Valonia.* Acorn cups of certain species of oak from South Europe, containing 25-35% of tannic acid.

*Vegetable alkali.* Potash.

*Verdigris.* Acetate of copper.

*Wet out, to.* To damp before putting the yarn or cloth into the dye.

# BIBLIOGRAPHY

A profitable Boke. (On Dyeing.) Translated from the Dutch. 1583.

Bancroft, Edward. The Philosophy of Permanent Colours. 1794.

Berthollet. The Art of Dyeing. 1824.

Bird, F. J. The Dyer's Handbook. 1875.

Bolton, Clement. A Manual of Wool Dyeing. 1913.

Boulger, Professor G. S. The Uses of plants. 1889.

Brewster's Edinburgh Encyclopædia. 1830. Dyeing.

Crook, W. Dyeing and Tissue Printing. 1882.

Darwin and Meldola. Woad. ("Nature," Nov. 12, 1896.)

Edge, Alfred. Some British Dye Lichens. (Journal of the Society of Dyers and Colourists. May, 1914.)

Edmonston, T. "On the Native Dyes of the Shetland Islands." (Transactions of Botanical Society of Edinburgh, Vol. 1, 1841.)

English Encyclopædia. Dyeing. 1802.

Francheville. On Ancient and Modern Dyes, 1767. (Royal Academy of Sciences, Berlin.)

Haigh, James. The Dyer's Assistant. 1778.

Hellot, Macquer, M. le Pilleur D' Apligny. The Art of Dyeing Wool, Silk and Cotton. (Translated from the French, 1789. New Edition. 1901.)

Henslow. Professor G. Use of British plants.

Hiscock, Gardiner D. 20th Century book of Recipes, Formulas and Processes. 1907.

Hummel, J. J. The Dyeing of Textile Fabrics.

Hurst, Silk Dyeing and Printing. (Technological Handbook. 1892.) [Pg 64]

Jarmain, George. On Wool Dyeing. 6 Lectures. 1876.

Knecht, Rawson and Lowenthal. A Manuel of Dyeing. 1893.

Lindsay, Dr. W. L. On the Dyeing Properties of Lichens. (Edinburgh New Philosophical Journal, 1855.)

Love, Thomas. The Practical Dyer and Scourer. 1854.

Mackay, Mrs. Anstruther. Simple Home Dyeing.

Milroy, R. P. Handbook for Dyeing for Woollen Homespun Workers. (Congested Districts Board for Ireland.)

Morris, William. "Of Dyeing as an Art." (Essays by Members of Arts and Crafts Exhibition Society, 1903.)

Morris, William. "The Lesser Arts of Life." (from Architecture, Industry and Wealth. 1902.)

Napier, James. A Manual of Dyeing Recipes. 1855.

Napier, James. A Manual of the Art of Dyeing. 1853.

Parnell's Applied Chemistry. — Article on Dyeing.

Plowright, Dr. British Dye Plants. (Journal of the Royal Horticultural Society, Vol. 26. 1901.)

Rawson, Gardiner and Laycock. A Dictionary of Dyes, Mordants. 1901.

Sansome. "Dyeing." 1888.

Sims, T. Dyeing and Bleaching. (British Manufacturing Industries. 1877.)

Smith, David. The Dyer's Instructor. 1847.

Smith. Practical Dyer's Guide. 1849.

Sowerby. English Botany.

Sowerby. Useful Plants of Great Britain.

The Art of Dyeing. (Translated from the German. 1705. Reprint, 1913.)

The Dyer and Colour Maker's Companion. 1859.

Thomson, John M. The Practical Dyer's Assistant. 1849.

[Pg 65]